Holt Science

Biology Video Labs on DVD
Lab Manual with Answer Key

HOLT, RINEHART AND WINSTON

A Harcourt Education Company

Orlando • **Austin** • New York • San Diego • Toronto • London

TO THE TEACHER

The Biology Video Labs on DVD allow your students to observe 43 biology labs. This accompanying lab manual allows your students to follow along with the procedures and write answers and conclusions directly on lab worksheets. The *Biology Video Labs on DVD Lab Manual with Answer Key* can be used in several ways to enhance science instruction. If space, time, and equipment are constraints for your classroom, the video labs and manual can be used to provide students with a virtual lab experience. The lab videos and manuals can also be used for "make-up work" by students who are absent during a class lab. The lab manual can even be used alone, with students performing the labs directly from the manual. In this case, you may wish to have students preview or refer to the video labs for procedural tips. Teacher Resource Pages for each lab can be found at the back of the manual.

ISBN 0-03-036752-2

1 2 3 4 5 6   862   08 07 06 05 04

Contents

Unit • Foundations of Biology

Unit • Cell Biology

Unit • Genetics and Biotechnology

Unit • Evolution

Unit • Ecology

Unit • Microbes, Protists, and Fungi

Unit • Human Biology

Lab Safety

In the laboratory, you can engage in hands-on explorations, test your scientific hypotheses, and build practical lab skills. However, while you are working in the lab or in the field, it is your responsibility to protect yourself and your classmates by conducting yourself in a safe manner. You will avoid accidents in the lab by following directions, handling materials carefully, and taking your work seriously. Read the following safety guidelines before working in the lab. Make sure that you understand all safety guidelines before entering the lab.

BEFORE YOU BEGIN

- **Read the entire activity before entering the lab.** Be familiar with the instructions before beginning an activity. Do not start an activity until you have asked your teacher to explain any parts of the activity that you do not understand.

- **Student-designed procedures or inquiry activities must be approved by your teacher before you attempt the procedures or activities.**

- **Wear the right clothing for lab work.** Before beginning work, tie back long hair, roll up loose sleeves, and put on any required personal protective equipment as directed by your teacher. Remove your wristwatch and any necklaces or jewelry that could get caught in moving parts. Avoid or confine loose clothing that could knock things over, catch on fire, get caught in moving parts, contact electrical connections, or absorb chemical solutions. Wear pants rather than shorts or skirts. Nylon and polyester fabrics burn and melt more readily than cotton does. Protect your feet from chemical spills and falling objects. Do not wear open-toed shoes, sandals, or canvas shoes in the lab. In addition, chemical fumes may react with and ruin some jewelry, such as pearl jewelry. Do not apply cosmetics in the lab. Some hair care products and nail polish are highly flammable.

- **Do not wear contact lenses in the lab.** Even though you will be wearing safety goggles, chemicals could get between contact lenses and your eyes and could cause irreparable eye damage. If your doctor requires that you wear contact lenses instead of glasses, then you should wear eye-cup safety goggles— similar to goggles worn for underwater swimming—in the lab. Ask your doctor or your teacher how to use eye-cup safety goggles to protect your eyes.

- **Know the location of <u>all safety and emergency equipment</u> used in the lab.** Know proper fire-drill procedures and the location of all fire exits. Ask your teacher where the nearest eyewash stations, safety blankets, safety shower, fire extinguisher, first-aid kit, and chemical spill kit are located. Be sure that you know how to operate the equipment safely.

WHILE YOU ARE WORKING

- **Always wear a lab apron and safety goggles.** Wear these items even if you are not working on an activity. Labs contain chemicals that can damage your clothing, skin, and eyes. If your safety goggles cloud up or are uncomfortable, ask your teacher for help. Lengthening the strap slightly, washing the goggles with soap and warm water, or using an anti-fog spray may help the problem.

- **NEVER work alone in the lab.** Work in the lab only when supervised by your teacher. Do not leave equipment unattended while it is in operation.

- **Perform only activities specifically assigned by your teacher.** Do not attempt any procedure without your teacher's direction. Use only materials and equipment listed in the activity or authorized by your teacher. Steps in a procedure should be performed only as described in the activity or as approved by your teacher.

- **Keep your work area neat and uncluttered.** Have only books and other materials that are needed to conduct the activity in the lab. Keep backpacks, purses, and other items in your desk, locker, or other designated storage areas.

- **Always heed safety symbols and cautions listed in activities, listed on handouts, posted in the room, provided on chemical labels, and given verbally by your teacher.** Be aware of the potential hazards of the required materials and procedures, and follow all precautions indicated.

- **Be alert, and walk with care in the lab.** Be aware of others near you and your equipment.

- **Do not take food, drinks, chewing gum, or tobacco products into the lab.** Do not store or eat food in the lab.

- **NEVER taste chemicals or allow them to contact your skin.** Keep your hands away from your face and mouth, even if you are wearing gloves.

- **Exercise caution when working with electrical equipment.** Do not use electrical equipment with frayed or twisted wires. Be sure that your hands are dry before using electrical equipment. Do not let electrical cords dangle from work stations. Dangling cords can cause you to trip and can cause an electrical shock.

- **Use extreme caution when working with hot plates and other heating devices.** Keep your head, hands, hair, and clothing away from the flame or heating area. Remember that metal surfaces connected to the heated area will become hot by conduction. Gas burners should be lit only with a spark lighter, not with matches. Make sure that all heating devices and gas valves are turned off before you leave the lab. Never leave a heating device unattended when it is in use. Metal, ceramic, and glass items do not necessarily look hot when they are hot. Allow all items to cool before storing them.

- **Do not fool around in the lab.** Take your lab work seriously, and behave appropriately in the lab. Lab equipment and apparatus are not toys; never use lab time or equipment for anything other than the intended purpose. Be aware of the safety of your classmates as well as your safety at all times.

EMERGENCY PROCEDURES

- **Follow standard fire-safety procedures.** If your clothing catches on fire, do not run; WALK to the safety shower, stand under it, and turn it on. While doing so, call to your teacher. In case of fire, alert your teacher and leave the lab.

- **Report any accident, incident, or hazard—no matter how trivial—to your teacher immediately.** Any incident involving bleeding, burns, fainting, nausea, dizziness, chemical exposure, or ingestion should also be reported immediately to the school nurse or to a physician. If you have a close call, tell your teacher so that you and your teacher can find a way to prevent it from happening again.

- **Report all spills to your teacher immediately.** Call your teacher rather than trying to clean a spill yourself. Your teacher will tell you whether it is safe for you to clean up the spill; if it is not safe, your teacher will know how to clean up the spill.

- **If you spill a chemical on your skin, wash the chemical off in the sink and call your teacher.** If you spill a solid chemical onto your clothing, brush it off carefully without scattering it onto somebody else and call your teacher. If you get liquid on your clothing, wash it off right away by using the faucet at the sink and call your teacher. If the spill is on your pants or something else that will not fit under the sink faucet, use the safety shower. Remove the pants or other affected clothing while you are under the shower, and call your teacher. (It may be temporarily embarrassing to remove pants or other clothing in front of your classmates, but failure to flush the chemical off your skin could cause permanent damage.)

WHEN YOU ARE FINISHED

- **Clean your work area at the conclusion of each lab period as directed by your teacher.** Broken glass, chemicals, and other waste products should be disposed of in separate, special containers. Dispose of waste materials as directed by your teacher. Put away all material and equipment according to your teacher's instructions. Report any damaged or missing equipment or materials to your teacher.

- **Wash your hands with soap and hot water after each lab period.** To avoid contamination, wash your hands at the conclusion of each lab period, and before you leave the lab.

Safety Symbols

Before you begin working in the lab, familiarize yourself with the following safety symbols, which are used throughout your textbook, and the guidelines that you should follow when you see these symbols.

 Eye Protection

- **Wear approved safety goggles as directed.** Safety goggles should be worn in the lab at all times, especially when you are working with a chemical or solution, a heat source, or a mechanical device.

- **If chemicals get into your eyes, flush your eyes immediately.** Go to an eyewash station immediately, and flush your eyes (including under the eyelids) with running water for at least 15 minutes. Use your thumb and fingers to hold your eyelids open and roll your eyeball around. While doing so, ask another student to notify your teacher.

- **Do not wear contact lenses in the lab.** Chemicals can be drawn up under a contact lens and into the eye. If you must wear contacts prescribed by a physician, tell your teacher. In this case, you must also wear approved eye-cup safety goggles to help protect your eyes.

- **Do not look directly at the sun or any light source through any optical device or lens system, and do not reflect direct sunlight to illuminate a microscope.** Such actions concentrate light rays to an intensity that can severely burn your retinas, which may cause blindness.

 Clothing Protection

- **Wear an apron or lab coat at all times in the lab to prevent chemicals or chemical solutions from contacting skin or clothes**.

- **Tie back long hair, secure loose clothing, and remove loose jewelry so that they do not knock over equipment, get caught in moving parts, or come into contact with hazardous materials.**

 Hand Safety

- **Do not cut an object while holding the object in your hand.** Dissect specimens in a dissecting tray.

- **Wear protective gloves when working with an open flame, chemicals, solutions, specimens, or live organisms.**

- **Use a hot mitt to handle resistors, light sources, and other equipment that may be hot.** Allow all equipment to cool before storing it.

Hygienic Care

- **Keep your hands away from your face and mouth while you are working in the lab.**

- **Wash your hands thoroughly before you leave the lab.**

- **Remove contaminated clothing immediately**. If you spill caustic substances on your skin or clothing, use the safety shower or a faucet to rinse. Remove affected clothing while you are under the shower, and call to your teacher. (It may be temporarily embarrassing to remove clothing in front of your classmates, but failure to rinse a chemical off your skin could result in permanent damage.)

- **Launder contaminated clothing separately.**

- **Use the proper technique demonstrated by your teacher when you are handling bacteria or other microorganisms.** Treat all microorganisms as if they are pathogens. Do not open Petri dishes to observe or count bacterial colonies.

- **Return all stock and experimental cultures to your teacher for proper disposal.**

Sharp-Object Safety

- **Use extreme care when handling all sharp and pointed instruments, such as scalpels, sharp probes, and knives.**

- **Do not cut an object while holding the object in your hand.** Cut objects on a suitable work surface. Always cut in a direction away from your body.

- **Do not use double-edged razor blades in the lab.**

Glassware Safety

- **Inspect glassware before use; do not use chipped or cracked glassware.** Use heat-resistant glassware for heating materials or storing hot liquids, and use tongs or a hot mitt to handle this equipment.

- **Do not attempt to insert glass tubing into a rubber stopper without specific instructions from your teacher.**

- **Notify immediately your teacher if a piece of glassware or a light bulb breaks.** Do not attempt to clean up broken glass unless your teacher directs you to do so.

Proper Waste Disposal

- Clean and sanitize all work surfaces and personal protective equipment after each lab period as directed by your teacher.

- Dispose of all sharp objects (such as broken glass) and other contaminated materials (biological or chemical) in special containers only as directed by your teacher. Never put these materials into a regular waste container or down the drain.

 Animal Care and Safety

- Do not approach or touch any wild animals. When working outdoors, be aware of poisonous or dangerous animals in the area.

- Always obtain your teacher's permission before bringing any animal (including a pet) into the school building.

- Handle animals only as directed by your teacher. Mishandling or abusing any animal will not be tolerated.

 Plant Safety

- Do not ingest any plant part used in the lab (especially commercially sold seeds). Do not touch any sap or plant juice directly. Always wear gloves.

- Wear disposable polyethylene gloves when handling any wild plant. Wash your hands thoroughly after handling any plant or plant part (particularly seeds). Avoid touching your face and eyes.

- Do not inhale or expose yourself to the smoke of any burning plant. Smoke contains irritants that can cause inflammation in the throat and lungs.

- Do not pick wildflowers or other plants unless directed to do so by your teacher.

 Electrical Safety

- Do not use equipment with frayed electrical cords or loose plugs.

- Fasten electrical cords to work surfaces by using tape. Doing so will prevent tripping and will ensure that equipment will not fall off the table.

- Do not use electrical equipment near water or when your clothing or hands are wet.

- Hold the rubber cord when you plug in or unplug equipment. Do not touch the metal prongs of the plug, and do not unplug equipment by pulling on the cord.

- Wire coils on hot plates may heat up rapidly. If heating occurs, open the switch immediately and use a hot mitt to handle the equipment.

Heating Safety

- Be aware of any source of flames, sparks, or heat (such as open flames, electric heating coils, or hot plates) before working with flammable liquids or gases.

- **Avoid using open flames.** If possible, work only with hot plates that have an on/off switch and an indicator light. Do not leave hot plates unattended. Do not use alcohol lamps. Turn off hot plates and open flames when they are not in use.

- Never leave a hot plate unattended while it is turned on or while it is cooling off.

- Know the location of lab fire extinguishers and fire-safety blankets.

- **Use tongs or appropriate insulated holders when handling heated objects.** Heated objects often do not appear to be hot. Do not pick up an object with your hand if it could be warm.

- Keep flammable substances away from heat, flames, and other ignition sources.

- Allow all equipment to cool before storing it.

 Fire Safety

- Know the location of lab fire extinguishers and fire-safety blankets.

- Know your school's fire-evacuation routes.

- If your clothing catches on fire, walk (do not run) to the emergency lab shower to put out the fire. If the shower is not working, STOP, DROP, and ROLL! Smother the fire by stopping immediately, dropping to the floor, and rolling until the fire is out.

 Safety with Gases

- **Do not inhale any gas or vapor unless directed to do so by your teacher.** Never inhale pure gases.

- **Handle materials that emit vapors or gases in a well-ventilated area.** This work should be done in an approved chemical fume hood.

 Caustic Substances

- If a chemical gets on your skin, on your clothing, or in your eyes, rinse it immediately and alert your teacher.

- If a chemical is spilled on the floor or lab bench, alert your teacher, but do not clean it up yourself unless your teacher directs you to do so.

Chemical Safety

- **Always wear safety goggles, gloves, and a lab apron or coat to protect your eyes and skin when you are working with any chemical or chemical solution.**

- **Do not taste, touch, or smell any chemicals or bring them close to your eyes unless specifically instructed to do so by your teacher.** If your teacher tells you to note the odor of a substance, do so by waving the fumes toward you with your hand. Do not pipette any chemicals by mouth; use a suction bulb as directed by your teacher.

- **Know where the emergency lab shower and eyewash stations are and how to use them.** If you get a chemical on your skin or clothing, wash it off at the sink while calling to your teacher.

- **Always handle chemicals or chemical solutions with care.** Check the labels on bottles, and observe safety procedures. Label beakers and test tubes containing chemicals.

- **For all chemicals, take only what you need.** Do not return unused chemicals or solutions to their original containers. Return unused reagent bottles or containers to your teacher.

- **NEVER take any chemicals out of the lab.**

- **Do not mix any chemicals unless specifically instructed to do so by your teacher.** Otherwise harmless chemicals can be poisonous or explosive if combined.

- **Do not pour water into a strong acid or base.** The mixture can produce heat and can splatter.

- **Report all spills to your teacher immediately.** Spills should be cleaned up promptly as directed by your teacher.

- **Do not allow radioactive materials to come into contact with your skin, hair, clothing, or personal belongings. Although the materials used in the lab are not hazardous when they are used properly, radioactive materials can cause serious illness and may have permanent effects if they are misused.**

Laboratory Techniques

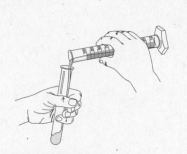

FIGURE A

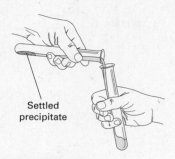

Settled
precipitate

FIGURE B

FIGURE C

HOW TO DECANT AND TRANSFER LIQUIDS

1. The safest way to transfer a liquid from a graduated cylinder to a test tube is shown in Figure A. The liquid is transferred at arm's length, with the elbows slightly bent. This position enables you to see what you are doing while maintaining steady control of the equipment.

2. Sometimes, liquids contain particles of insoluble solids that sink to the bottom of a test tube or beaker. Use one of the methods shown above to separate a supernatant (the clear fluid) from insoluble solids.

   a. Figure B shows the proper method of decanting a supernatant liquid from a test tube.

   b. Figure C shows the proper method of decanting a supernatant liquid from a beaker by using a stirring rod. The rod should touch the wall of the receiving container. Hold the stirring rod against the lip of the beaker containing the supernatant. As you pour, the liquid will run down the rod and fall into the beaker resting below. When you use this method, the liquid will not run down the side of the beaker from which you are pouring.

HOW TO HEAT SUBSTANCES AND EVAPORATE SOLUTIONS

1. Use care in selecting glassware for high-temperature heating. The glassware should be heat resistant.

2. When heating glassware by using a gas flame, use a ceramic-centered wire gauze to protect glassware from direct contact with the flame. Wire gauzes can withstand extremely high temperatures and will help prevent glassware from breaking. Figure D shows the proper setup for evaporating a solution over a water bath.

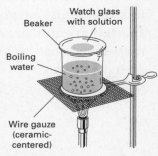

Beaker

Watch glass
with solution

Boiling
water

Wire gauze
(ceramic-
centered)

FIGURE D

3. In some experiments, you are required to heat a substance to high temperatures in a porcelain crucible. Figure E shows the proper apparatus setup used to accomplish this task.

4. Figure F shows the proper setup for evaporating a solution in a porcelain evaporating dish with a watch glass cover that prevents spattering.

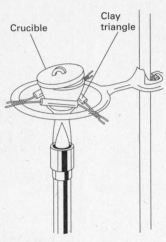

FIGURE E

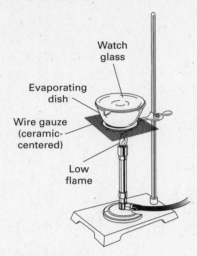

FIGURE F

5. Glassware, porcelain, and iron rings that have been heated may *look* cool after they are removed from a heat source, but these items can still burn your skin even after several minutes of cooling. Use tongs, test-tube holders, or heat-resistant mitts and pads whenever you handle these pieces of apparatus.

6. You can test the temperature of beakers, ring stands, wire gauzes, or other pieces of apparatus that have been heated by holding the back of your hand close to their surfaces before grasping them. You will be able to feel any energy as heat generated from the hot surfaces. DO NOT TOUCH THE APPARATUS. Allow plenty of time for the apparatus to cool before handling.

HOW TO POUR LIQUID FROM A REAGENT BOTTLE

1. Read the label at least three times before using the contents of a reagent bottle.

2. Never lay the stopper of a reagent bottle on the lab table.

3. When pouring a caustic or corrosive liquid into a beaker, use a stirring rod to avoid drips and spills. Hold the stirring rod against the lip of the reagent bottle. Estimate the amount of liquid you need, and pour this amount along the rod, into the beaker. See Figure G.

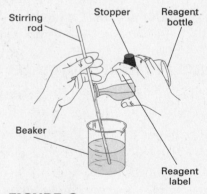

FIGURE G

4. Extra precaution should be taken when handling a bottle of acid. Remember the following important rules: Never add water to any concentrated acid, particularly sulfuric acid, because the mixture can splash and will generate a lot of energy as heat. To dilute any acid, add the acid to water in small quantities while stirring slowly. Remember the "triple A's"—*Always Add Acid* to water.

5. Examine the outside of the reagent bottle for any liquid that has dripped down the bottle or spilled on the counter top. Your teacher will show you the proper procedures for cleaning up a chemical spill.

6. Never pour reagents back into stock bottles. At the end of the experiment, your teacher will tell you how to dispose of any excess chemicals.

HOW TO HEAT MATERIAL IN A TEST TUBE

1. Check to see that the test tube is heat resistant.

2. Always use a test tube holder or clamp when heating a test tube.

3. Never point a heated test tube at anyone, because the liquid may splash out of the test tube.

4. Never look down into the test tube while heating it.

5. Heat the test tube from the upper portions of the tube downward, and continuously move the test tube, as shown in Figure H. Do not heat any one spot on the test tube. Otherwise, a pressure buildup may cause the bottom of the tube to blow out.

HOW TO USE A MORTAR AND PESTLE

1. A mortar and pestle should be used for grinding only one substance at a time. See Figure I.

2. Never use a mortar and pestle for simultaneously mixing different substances.

3. Place the substance to be broken up into the mortar.

4. Pound the substance with the pestle, and grind to pulverize.

5. Remove the powdered substance with a porcelain spoon.

HOW TO DETECT ODORS SAFELY

1. Test for the odor of gases by wafting your hand over the test tube and cautiously sniffing the fumes as shown in Figure J.

2. Do not inhale any fumes directly.

3. Use a fume hood whenever poisonous or irritating fumes are present. DO NOT waft and sniff poisonous or irritating fumes.

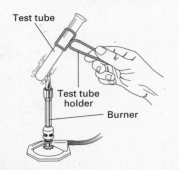

FIGURE H

FIGURE I

FIGURE J

Safety in the Field

Activities conducted outdoors require some advance planning to ensure a safe environment. The following general guidelines should be followed for fieldwork.

- **Know your mission.** Your teacher will tell you the goal of the field trip in advance. Be sure to have your permission slip approved before the trip, and check to be sure that you have all necessary supplies for the day's activity.

- **Find out about on-site hazards before setting out.** Determine whether poisonous plants or dangerous animals are likely to be present where you are going. Know how to identify these hazards. Find out about other hazards, such as steep or slippery terrain.

- **Wear protective clothing.** Dress in a manner that will keep you warm, comfortable, and dry. Decide in advance whether you will need sunglasses, a hat, gloves, boots, or rain gear to suit the terrain and local weather conditions.

- **Do not approach or touch wild animals.** If you see a threatening animal, call your teacher immediately. Avoid any living thing that may sting, bite, scratch, or otherwise cause injury.

- **Do not touch wild plants or pick wildflowers unless specifically instructed to do so by your teacher.** Many wild plants can be irritating or toxic. Never taste any wild plant.

- **Do not wander away from others.** Travel with a partner at all times. Stay within an area where you can be seen or heard in case you run into trouble.

- **Report all hazards or accidents to your teacher immediately.** Even if the incident seems unimportant, let your teacher know what happened.

- **Maintain the safety of the environment.** Do not remove anything from the field site without your teacher's permission. Stay on trails, when possible, to avoid trampling delicate vegetation. Never leave garbage behind at a field site. Leave natural areas as you found them.

Safety with Organisms and Biological Specimens

Consider these guidelines before using living organisms or biological specimens in the laboratory:

BE AWARE OF INDIVIDUAL HEALTH RISKS

Do not allow students with open cuts, abrasions, or sores to work with live organisms or biological specimens. Consult with the school nurse to screen students who may be exceptionally vulnerable to allergic reactions or microbial infections. These students should not participate in laboratory activities without written permission from a physician.

BE AWARE OF DANGEROUS ORGANISMS

In the school laboratory, you should treat the following items as potentially pathogenic (disease-causing): all microorganisms; microbial culture media, whether used or unused; and biological specimens, including dead organisms, parts of organisms, bodily fluids, and food items.

Known pathogens are not appropriate for use in school laboratories. Do not use blood agar culture media, and do not cultivate microbes from a human or animal source.

Both living and dead specimens may present additional hazards, such as sharp objects and hazardous chemicals. Investigate the potential hazards of each organism or specimen and alert students to these hazards.

PRACTICE ASEPTIC TECHNIQUE

Students should never taste or eat anything in the laboratory and should avoid touching their faces while working in the laboratory. Students should wear goggles, aprons, and gloves whenever working with organisms or specimens.

Prior to conducting a lab activity involving microbes, demonstrate correct aseptic technique to students. Never pipet liquid media by mouth. Use sterile cotton applicator sticks in place of inoculating loops and Bunsen burners for inoculating microbial cultures. Seal with tape all Petri dishes containing bacterial cultures.

HANDLE ANIMALS RESPONSIBLY

Insist that students treat live animals carefully and humanely in order to minimize danger to themselves and to the animals. Be alert to the potential for allergies, bites, scratches, or stings. The National Association of Biology Teachers has endorsed the "Principles and Guidelines for the Use of Animals in Precollege Education," available from the Institute of Laboratory Animals Resources (ILAR), part of the National Research Council, at 500 Fifth Street NW, Washington, DC 20001, phone 202 334-2590, e-mail ILAR@nas.edu, and http://dels.nas.edu/ilar.

ENSURE PROPER CLEANUP AND DISPOSAL

Have procedures and materials prepared for cleanup of spills and accidents involving biological materials as well as hazardous chemicals. Lab surfaces and large equipment should be wiped with a disinfectant cleaning solution before and after activities. Perform cleanup of hazardous materials yourself—do not allow students to do this.

The following should be autoclaved or sterilized before disposal: microbial cultures and consumable or disposable culture media; excretions or residues from live organisms; consumable or disposable materials (such as paper, soil, or liquids) that have come in contact with living organisms or microbial cultures; and any handheld equipment that has come in contact with living organisms or microbial culture media during a lab.

Autoclave or steam-sterilize items at 120°C and 15 psi for 15–20 minutes. If sterilizing devices are not available, soak items in household bleach (full strength) for 30 minutes and then allow to dry.

Student Safety Contract

Read carefully the Student Safety Contract below. Then, fill in your name in the first blank, date the contract, and sign it.

Student Safety Contract

I will

- read the lab investigation before coming to class
- wear personal protective equipment as directed to protect my eyes, face, hands, and body while conducting class activities
- follow all instructions given by the teacher
- conduct myself in a responsible manner at all times in a laboratory situation

I, _____, have read and agree to abide by the safety regulations as set forth above and any additional printed instructions provided by my teacher or the school district.

I agree to follow all other written and oral instructions given in class.

Date: _____

Signature: _____

Ordering Lab Materials

ORDERING LAB MATERIALS WITH THE
ONE-STOP PLANNER® CD-ROM

Your class and prep time are valuable. Now, it's easier and faster than ever to organize and obtain the materials that you need for all of the labs in this lab program. Using the Lab Materials QuickList software found on the *One-Stop Planner® CD-ROM*, you can do the following:

• View all of the materials that you need for any (or all) labs.

The Lab Materials QuickList software allows you to easily see all of the materials needed for any lab in this lab program. Use your materials list to order all of your materials at once. Or use the list to determine what items you need to resupply or supplement your stockroom so that you'll be prepared to do any lab.

• Create a customized materials list.

No two teachers teach exactly alike. The Lab Materials QuickList software allows you to select labs by type of lab or by type of material used (such as a chemical or living organism). You can create a materials list that summarizes the needs of whichever labs you choose.

• Let the software handle the details.

You can customize your list based on the number of students and the number of lab groups. A powerful software engine that has been programmed to distinguish between consumable and nonconsumable materials will "do the math." Whether you're examining all of the labs for a whole year or just the labs that you're planning for next week, the software does the hard work of totaling and tallying, telling you what you need and *exactly* how much you'll need for the labs that you've selected.

• Print your list.

By printing out materials lists that you created by using the Lab Materials QuickList software, you can have a copy of any materials list right at your fingertips for easy reference at any time.

• Order your materials easily.

After you've created your materials list by using the Lab Materials QuickList software, you can use it to order from WARD'S or to prepare a purchase order to be sent directly to another scientific materials supplier.

Visit go.hrw.com to learn more about the Lab Materials QuickList software.

Laboratory Assessment

As a teacher, only you know the best assessment methods to apply to student lab work in your classes.

After each lab, you may want students to prepare a lab report. A traditional lab report usually includes at least the following components:

- **title**
- **summary paragraph** describing the purpose and procedure
- **data tables** and **observations** that are organized and comprehensive
- **answers** to the Analysis and Conclusions questions

One-Stop Planner CD-ROM *Provides Scoring Rubrics*

The *One-Stop Planner*® *CD-ROM* includes *scoring rubrics* to expand your assessment options. These assessment tools can provide a means to objective assessment or serve as a good starting point for designing your own specialized assessment tools. Laboratory-related rubrics and checklists include the following:

- Scoring Rubric for Skills Practice Labs
- Scoring Rubrics for Lab Techniques and Experimental Design Labs
- Scoring Rubrics for Biotechnology Labs
- Scoring Rubric for Inquiry Labs

The *One-Stop Planner*® *CD-ROM* also includes the following resources that can help you emphasize the importance of safe lab behavior:

- Student Safety Contracts
- Safety Test

Name _____ Class _____ Date _____

Observing the Effects of Acid Rain on Seeds

SKILLS

- Using scientific methods
- Collecting, organizing, and graphing data

OBJECTIVES

- **Use** a scientific method to investigate a problem.
- **Predict** how acid rain affects germination and growth.

MATERIALS

- safety goggles
- protective gloves
- lab apron
- 50 seeds
- 250 mL beaker
- 20 mL mold inhibitor
- distilled water
- paper towels
- solutions of different pH
- wax pencil or marker
- resealable plastic bags
- metric ruler
- graph paper

SAFETY

 CAUTION: Always wear safety goggles and a lab apron to protect your eyes and clothing.

 CAUTION: Do not touch or taste any chemicals. Know the location of the emergency shower and eyewash station and how to use them. If you get a chemical on your skin or clothing, wash it off at the sink while calling to the teacher. Notify the teacher of a spill. Spills should be cleaned up promptly, according to your teacher's directions.

 CAUTION: Glassware is fragile. Notify the teacher of broken glass or cuts. Do not clean up broken glass or spills with broken glass unless the teacher tells you to do so.

Before You Begin

Living things, such as salamander embryos, can be damaged by **acid rain** at certain times during their lives. In this lab, you will investigate the effect of acidic solutions on seeds. One way to investigate a problem is to design and conduct an **experiment.** We begin a scientific investigation by making **observations** and asking questions.

1. Write a definition for each boldface term in the paragraph above and for each of the following terms: pH, hypothesis, prediction, variable, control group. Use a separate sheet of paper.

2. Based on the objectives for this lab, write a question you would like to explore about the effect of acid rain, for example, When is a plant most susceptible to acid rain?

_____

_____

Procedure
PART A: DESIGN AN EXPERIMENT

1. Work with members of your lab group to explore one of the questions written for step 2 of **Before You Begin.** To explore the question, design an experiment that uses the materials listed for this lab.

> **You Choose**
> As you design your experiment, decide the following:
> **a.** what question you will explore
> **b.** what hypothesis you will test
> **c.** how to simulate growing seeds in soil moistened by acid rain
> **d.** how to keep seeds moist during the experiment
> **e.** what your test solutions and control will be
> **f.** how to measure seedling growth
> **g.** what to record in your data table

2. Write a procedure for your experiment. Make a list of all the safety precautions you will take. Have your teacher approve your procedure and safety precautions before you begin the experiment.

Observing the Effects of Acid Rain on Seeds *continued*

PART B: CONDUCT YOUR EXPERIMENT

3. Put on safety goggles, protective gloves, and a lab apron.

4. Place your seeds in a 250 mL beaker, and slowly add enough mold inhibitor to cover the seeds. **CAUTION: The mold inhibitor contains household bleach, which is a base.** Soak the seeds for 10 minutes, and then pour the mold inhibitor into the proper waste container. Gently rinse the seeds with distilled water, and place them on clean paper towels.

5. Set up your group's experiment. **CAUTION: Solutions with a pH below 7.0 are acids.** Conduct your experiment for 7–10 days. Make observations every 1–2 days, and note any changes. Record each day's observations in the data table below.

DATA TABLE		
Solution	**Date**	**Observations**

PART C: CLEANUP AND DISPOSAL

6. Dispose of solutions, broken glass, and seeds in the designated waste containers. Do not pour chemicals down the drain or put lab materials in the trash unless your teacher tells you to do so.

7. Clean up your work area and all lab equipment. Return lab equipment to its proper place. Wash your hands thoroughly before you leave the lab and after you finish all work.

ANALYZE AND CONCLUDE

1. Summarizing Results Describe any changes in the look of your seeds during the experiment. Discuss seed type, average seed size, number of germinated seeds, and changes in seedling length.

_____

_____

Observing the Effects of Acid Rain on Seeds *continued*

2. Analyzing Results Were there any differences between the solutions? Explain.

_____

_____

3. Analyzing Methods What was the control group in your experiment?

_____

4. Analyzing Data Use graph paper to make graphs of your group's data. Plot seedling growth (in millimeters) on the *y*-axis. Plot number of days on the *x*-axis.

5. Relating Concepts What scientific methods did you use to design and conduct your experiment?

_____

_____

_____

6. Evaluating Methods How could your experiment be improved?

_____

_____

7. Inferring Conclusions How do acidic conditions appear to affect seeds?

_____

_____

8. Predicting Outcomes How might acid rain affect the plants in an ecosystem?

_____

_____

9. Further Inquiry Write a new question about the effect of acid rain that could be explored with another investigation.

_____

_____

Exploration Lab

Observing Enzyme Detergents

SKILLS

- Using scientific methods
- Measuring volume, mass, and pH

OBJECTIVES

- **Recognize** the function of enzymes in laundry detergents.
- **Relate** temperature and pH to the activity of enzymes.

MATERIALS

- safety goggles and lab apron
- balance
- graduated cylinder
- glass stirring rod
- 150 mL beaker
- 18 g regular instant gelatin or 1.8 g sugar-free instant gelatin
- 0.7 g Na_2CO_3
- tongs or a hot mitt
- 50 mL boiling water
- thermometer
- pH paper
- 6 test tubes
- test-tube rack
- pipet with bulb
- plastic wrap
- tape
- 50 mL beakers (6)
- 50 mL distilled water
- 1 g each of 5 brands of laundry detergent
- wax pencil
- metric ruler

SAFETY

 CAUTION: Always wear safety goggles and a lab apron to protect your eyes and clothing.

CAUTION: Do not touch or taste any chemicals. Know the location of the emergency shower and eyewash station and how to use them. If you get a chemical on your skin or clothing, wash it off at the sink while calling to the teacher. Notify the teacher of a spill. Spills should be cleaned up promptly, according to your teacher's directions.

CAUTION: Glassware is fragile. Notify the teacher of broken glass or cuts. Do not clean up broken glass or spills with broken glass unless the teacher tells you to do so.

Before You Begin

Enzymes are substances that speed up chemical reactions. Each enzyme operates best at a particular **pH** and temperature. Substances on which enzymes act are called **substrates.** Many enzymes are named for their substrates. For example, a **protease** is an enzyme that helps break down proteins. In this lab, you will investigate the effectiveness of laundry detergents that contain enzymes.

1. Write a definition for each boldface term in the paragraph above. Use a separate sheet of paper.

2. Based on the objectives for this lab, write a question you would like to explore about enzyme detergents.

_____

_____

Procedure
PART A: MAKE A PROTEIN SUBSTRATE

1. Put on safety goggles and a lab apron.

2. **CAUTION: Use tongs or a hot mitt to handle heated glassware.** Put 18 g of regular (1.8 g of sugar-free) instant gelatin in a 150 mL beaker. Slowly add 50 mL of boiling water to the beaker, and stir the mixture with a stirring rod. Test and record the pH of this solution.

3. Very slowly add 0.7 g of Na_2CO_3 to the hot gelatin while stirring. Note any reaction. Test and record the pH of this solution.

4. Place 6 test tubes in a test-tube rack. Pour 5 mL of the gelatin-Na_2CO_3 mixture into each tube. Use a pipet to remove any bubbles from the surface of the mixture in each tube. Cover the tubes tightly with plastic wrap and tape.

Observing Enzyme Detergents *continued*

Cool the tubes, and store them at room temperature until you begin Part C. Complete step 12.

PART B: DESIGN AN EXPERIMENT

5. Work with members of your lab group to explore one of the questions written for step 2 of **Before You Begin.** To explore the question, design an experiment that uses the materials listed for this lab.

You Choose

As you design your experiment, decide the following:

a. what question you will explore

b. what hypothesis you will test

c. what detergent samples you will test

d. what your control will be

e. how much of each solution to use for each test

f. how to determine if protein is breaking down

g. what data to record in your data table

---

6. Write a procedure for your experiment. Make a list of all the safety precautions you will take. Have your teacher approve your procedure and safety precautions before you begin the experiment.

PART C: CONDUCT YOUR EXPERIMENT

7. Put on safety goggles and a lab apron.

8. Make a 10 percent solution of each laundry detergent by dissolving 1 g of detergent in 9 mL of distilled water.

9. Set up your experiment. Repeat step 12.

10. Record your data after 24 hours.

PART D: CLEANUP AND DISPOSAL

11. Dispose of solutions, broken glass, and gelatin in the designated waste containers. Do not pour chemicals down the drain or put lab materials in the trash unless your teacher tells you to do so.

12. Clean up your work area and all lab equipment. Return lab equipment to its proper place. Wash your hands thoroughly before leaving the lab and after finishing all work.

| Observing Enzyme Detergents *continued*

Analyze and Conclude

1. Analyzing Methods Suggest a reason for adding Na$_2$CO$_3$ to the gelatin solution.

_____

_____

2. Analyzing Results Make a bar graph of your data. Plot the amount of gelatin broken down (change in the depth of the gelatin) on the *y*-axis and detergent on the *x*-axis. Use a separate sheet of graph paper.

3. Inferring Conclusions What conclusions did your group infer from the results? Explain.

_____

_____

4. Further Inquiry Write a new question about enzyme detergents that could be explored with another investigation.

_____

_____

Skills Practice Lab

Studying Animal Cells and Plant Cells

SKILLS

- Using a compound microscope
- Drawing

OBJECTIVES

- **Identify** the structures you can see in animal cells and plant cells.
- **Compare** and **contrast** the structure of animal cells and plant cells.

MATERIALS

- compound light microscope
- prepared slide of human epithelial cells
- safety goggles
- lab apron
- polyethylene gloves
- sprig of *Elodea*
- forceps
- microscope slides and cover slips
- dropper bottle of Lugol's iodine solution

SAFETY

 CAUTION: Always wear safety goggles and a lab apron to protect your eyes and clothing.

CAUTION: Do not touch or taste any chemicals. Know the location of the emergency shower and eyewash station and how to use them. If you get a chemical on your skin or clothing, wash it off at the sink while calling the teacher. Notify the teacher of a spill. Spills should be cleaned up promptly, according to your teacher's directions.

CAUTION: Glassware is fragile. Notify the teacher of broken glass or cuts. Do not clean up broken glass or spills with broken glass unless the teacher tells you to do so.

Before You Begin

You can see many cell parts with a **light microscope.** In animal cells, the **cytoplasm, cell membrane, nucleus, nucleolus,** and **vacuoles** can be seen. In plants cells, the **cell wall** and **chloroplasts** can also be be seen. Stains add color to cell parts and make them more visible with a light microscope. A stain can even make the **endoplasmic reticulum** visible. In this lab, you will use a light microscope to examine animal and plant cells.

Studying Animal Cells and Plant Cells *continued*

1. Write a definition for each boldface term in the paragraph above. Use a separate sheet of paper.

2. Why might a stain be needed to see cell parts under a microscope?

_____

_____

_____

3. Based on the objectives for this lab, write a question you would like to explore about cell structure.

_____

_____

Procedure

PART A: ANIMAL CELLS

1. 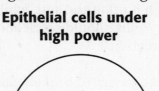 Examine a prepared slide of human epithelial cells under low power with a compound light microscope. Find cells that are separate from each other, and place them in the center of the field of view. Switch to high power, and adjust the diaphragm until you can see the cells more clearly. Identify as many cell parts as you can. *Note: Remember to use only the fine adjustment to focus at high power.*

2. Draw two or three epithelial cells as they look under high power. Label the cell membrane, the cytoplasm, the nuclear envelope, and the nucleus of at least one of the cells. Make a second drawing of these cells as you imagine they might look in the lining of your mouth.

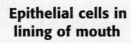

Epithelial cells under high power	Epithelial cells in lining of mouth

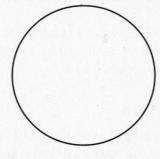

Name _____ Class _____ Date _____

Studying Animal Cells and Plant Cells *continued*

PART B: PLANT CELLS

3. 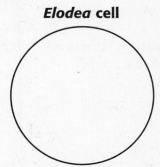 Using forceps, carefully remove a small leaf from near the top of an *Elodea* sprig. Place the whole leaf in a drop of water on a slide, and add a cover slip.

4. Observe the leaf under low power. Look for an area of the leaf in which you can see the cells clearly, and move the slide so that this area is in the center of the field of view. Switch to high power, and, if necessary, adjust the diaphragm. Identify as many cell parts as you can.

5. Find an *Elodea* cell in which you can see the chloroplasts clearly. Draw this cell. Label the cell wall, a chloroplast, and any other cell parts that you can see.

Elodea cell

6. Notice if the chloroplasts are moving in any of the cells. If you do not see movement, warm the slide in your hand or under a bright lamp for a minute or two. Look for movement of the cell contents again under the high power. Such movement is called cytoplasmic streaming.

7. 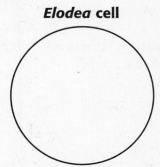 Put on safety goggles and a lab apron. Make a wet mount of another *Elodea* leaf, using Lugol's iodine solution instead of water. **CAUTION: Lugol's solution stains skin and clothing. Promptly wash off spills.** Observe these cells under low and high power.

8. Draw a stained *Elodea* cell. Label the cell wall and a chloroplast, as well as the central vacuole, the nucleus, and the cell membrane if they are visible.

Stained Elodea cell

PART C: CLEANUP AND DISPOSAL

9. 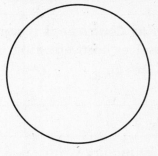 Dispose of solutions, broken glass, and *Elodea* leaves in the waste containers designated by your teacher. Do not pour chemicals down the drain or put lab materials in the trash unless your teacher tells you to do so.

10. Clean up your work area and all lab equipment. Return lab equipment to its proper place. Wash your hands thoroughly before you leave the lab and after you finish all work.

Studying Animal Cells and Plant Cells *continued*

Analyze and Conclude

1. Recognizing Patterns In what observable ways are animal and plant cells similar in structure, and in what observable ways are they different?

_____

_____

_____

_____

2. Comparing Structures Compare and contrast the cytoplasm of epithelial cells and *Elodea* cells.

_____

_____

_____

_____

3. Analyzing Methods What is the reason for staining *Elodea* cells with iodine?

_____

_____

4. Inferring Conclusions Lugol's iodine solution causes the movement of chloroplasts to stop. Explain why.

_____

_____

5. Inferring Conclusions If some of the epithelial cells were folded over on themselves but were still transparent, what could you conclude about their thickness?

_____

_____

6. Further Inquiry Write a new question about cell structure that could be explored with another investigation.

_____

_____

_____

Exploration Lab

Analyzing the Effect of Cell Size on Diffusion

SKILLS

- Using scientific methods
- Collecting, organizing, and graphing data

OBJECTIVES

- **Relate** the size of a cell to its surface area-to-volume ratio.
- **Predict** how the surface area-to-volume ratio of a cell will affect the diffusion of substances into the cell.

MATERIALS

- safety goggles
- lab apron
- disposable gloves
- block of phenolphthalein agar (3 × 3 × 6 cm)
- plastic knife
- metric ruler
- 250 mL beaker
- 150 mL of vinegar
- plastic spoon
- paper towel

SAFETY

 CAUTION: Always wear safety goggles and a lab apron to protect your eyes and clothing.

CAUTION: Do not touch or taste any chemicals. Know the location of the emergency shower and eyewash station and how to use them. If you get a chemical on your skin or clothing, wash it off at the sink while calling to the teacher. Notify the teacher of a spill. Spills should be cleaned up promptly, according to your teacher's directions.

CAUTION: Glassware is fragile. Notify the teacher of broken glass or cuts. Do not clean up broken glass or spills with broken glass unless the teacher tells you to do so.

| Analyzing the Effect of Cell Size on Diffusion *continued*

Before You Begin

Substances enter and leave a cell in several ways, including by **diffusion.** How efficiently a cell can exchange substances depends on the **surface area-to-volume ratio** (surface area ÷ volume) of the cell. **Surface area** is the size of the outside of an object. **Volume** is the amount of space an object takes up. In this lab, you will investigate how cell size affects the diffusion of substances into a cell. To do this, you will make cell models using agar that contains an indicator. This indicator will change color when an acidic solution diffuses into it.

1. Write a definition for each boldface term in the paragraph above. Use a separate sheet of paper.

2. Based on the objectives for this lab, write a question you would like to explore about cell size and diffusion.

_____

_____

Procedure

PART A: DESIGN AN EXPERIMENT

1. Work with members of your lab group to explore one of the questions written for step 2 of **Before You Begin.** To explore the question, design an experiment that uses the materials listed for this lab.

> **You Choose**
>
> As you design your experiment, decide the following:
>
>   **a.** what question you will explore
>
>   **b.** what hypothesis you will test
>
>   **c.** how many "cells" (agar cubes) you will have and what sizes they will be
>
>   **d.** how long to leave the "cells" in the vinegar
>
>   **e.** how to determine how far the vinegar diffused into a "cell"
>
>   **f.** how to prevent contamination of agar cubes as you handle them
>
>   **g.** what data to record in your data table

2. Write a procedure for your experiment. Make a list of all the safety precautions you will take. Have your teacher approve your procedure and safety precautions before you begin the experiment.

PART B: CONDUCT YOUR EXPERIMENT

3. ◆ ◆ ◆ Put on safety goggles, a lab apron, and disposable gloves.

4. ◆ ◆ ◆ Carry out the experiment you designed. Record your observations in your data table.

PART C: CLEANUP AND DISPOSAL

5. Dispose of solutions, broken glass, and agar in the designated waste containers. Do not pour chemicals down the drain or put lab materials in the trash unless your teacher tells you to do so.

6. Clean up your work area and all lab equipment. Return lab equipment to its proper place. Wash your hands thoroughly before you leave the lab and after you finish all work.

Analyze and Conclude

1. **Summarizing Results** Describe any changes in the appearance of the cubes.

_____

_____

2. **Summarizing Results** Make a graph using your group's data. Plot "Diffusion Distance (mm)" on the vertical axis. Use graph paper to plot "Surface Area-to-Volume Ratio" on the horizontal axis.

_____

_____

3. **Analyzing Results** Using the graph you made in item 2, make a statement about the relationship between the surface area-to-volume ratio and the distance a substance diffuses.

_____

_____

4. **Summarizing Results** Make a graph using your group's data. Use graph paper to plot "Rate of Diffusion (mm/min)" (distance vinegar moved ÷ time) on the vertical axis. Plot "Surface Area-to-Volume Ratio" on the horizontal axis.

_____

_____

5. **Analyzing Results** Using the graph you made in item 4, make a statement about the relationship between the surface area-to-volume ratio and the rate of diffusion of a substance.

_____

_____

Analyzing the Effect of Cell Size on Diffusion *continued*

6. Evaluating Methods In what ways do your agar models simplify or fail to simulate the features of real cells?

_____

_____

_____

7. Calculating Calculate the surface area and volume of a cube with a side length of 5 cm. Calculate the surface area and volume of a cube with a side length of 10 cm. Determine the surface area-to-volume ratio of each of these cubes. Which cube has the greater surface area-to-volume ratio?

_____

_____

_____

_____

8. Evaluating Conclusions How does the size of a cell affect the diffusion of substances into the cell?

_____

_____

_____

9. Further Inquiry Write a new question about cell size and diffusion that could be explored with another investigation.

_____

_____

Skills Practice Lab

Observing Oxygen Production from Photosynthesis

SKILLS

- Measuring
- Collecting Data
- Graphing

OBJECTIVE

- **Measure** amount of oxygen produced by an *Elodea* sprig.

MATERIALS

- 500 mL of 5 percent baking-soda-and-water solution
- 600 mL beaker
- 20 cm long *Elodea* sprigs (2–3)
- glass funnel
- test tube
- metric ruler
- protective gloves

SAFETY

 CAUTION: Always wear safety goggles and a lab apron to protect your eyes and clothing.

 CAUTION: Glassware is fragile. Notify the teacher of broken glass or cuts. Do not clean up broken glass or spills with broken glass unless the teacher tells you to do so.

 CAUTION: Wear disposable polyethlene gloves when handling any plant. Do not eat any part of a plant or plant seed used in the lab. Wash hands thoroughly after handling any part of a plant.

Before You Begin

Plants use **photosynthesis** to produce food. One product of photosynthesis is oxygen. In this activity, you will observe the process of photosynthesis and determine the rate of photosynthesis for *Elodea*.

1. Write a definition for the boldface term above.

_____

_____

2. You will be using the data table provided to record your data.

Procedure

1. Add 450 mL of baking-soda-and-water solution to a beaker.

2. Put two or three sprigs of *Elodea* in the beaker. The baking soda will provide the *Elodea* with the carbon dioxide it needs for photosynthesis.

3. Place the wide end of the funnel over the *Elodea*. The end of the funnel with the small opening should be pointing up. The *Elodea* and the funnel should be completely under the solution.

4. Fill a test tube with the remaining baking-soda-and-water solution. Place your thumb over the end of the test tube. Turn the test tube upside-down, taking care that no air enters. Hold the opening of the test tube under the solution and place the test tube over the small end of the funnel. Try not to let any solution leak out of the test tube as you do this.

5. Place the beaker setup in a well-lit area near a lamp or in direct sunlight.

6. Record that there was 0 mm gas in the test tube on day 0. (If you were unable to place the test tube without getting air in the tube, measure the height of the column of air in the test tube in millimeters. Record this value for day 0.) In this lab, change in gas volume is indicated by a linear measurement expressed in millimeters.

Data Table		
Amount of Gas Present in the Test Tube		
Days of exposure to light	**Total amount of gas present (mm)**	**Amount of gas produced per day (mm)**
0		
1		
2		
3		
4		
5		

7. For days 1 through 5, measure the amount of gas in the test tube. Record the measurements in your data table under the heading, "Total amount of gas present (mm)."

8. Calculate the amount of gas produced each day by subtracting the amount of gas present on the previous day from the amount of gas present today. Record these amounts under the heading, "Amount of gas produced per day (mm)."

9. Plot the data from your table on a graph.

Name _____ Class _____ Date _____

Observing Oxygen Production from Photosynthesis *continued*

Analyze and Conclude

1. **Summarizing Results** Using information from your graph, describe what happened to the amount of gas in the test tube.

_____

_____

2. **Analyzing Data** How much gas was produced in the test tube after day 5?

_____

_____

_____

3. **Drawing Conclusions** Write the equation for photosynthesis. Explain each part of the equation. For example, what ingredients are necessary for photosynthesis to take place? What substances are produced by photosynthesis? What gas is produced that we need in order to live?

_____

_____

_____

_____

_____

_____

4. **Predicting Patterns** What may happen to the oxygen level if an animal, such as a snail, were put in the beaker with the *Elodea* sprig while the *Elodea* sprig was making oxygen?

_____

_____

5. **Further Inquiry** Write a new question about photosynthesis that could be explored with another investigation.

_____

_____

Name _____ Class _____ Date _____

Modeling Mitosis

SKILLS
- Modeling
- Using scientific methods

OBJECTIVES
- **Describe** the events that occur in each stage of mitosis.
- **Relate** mitosis to genetic continuity.

MATERIALS
- pipe cleaners of at least two different colors
- yarn
- wooden beads
- white labels
- scissors

Before You Begin

The cell cycle includes all of the phases in the life of a cell. The **cell cycle** is a repeating sequence of cellular growth and division during the life of an organism. Mitosis is one of the phases in the cell cycle. **Mitosis** is the process by which the material in a cell's nucleus is divided during cell reproduction. In this lab, you will build a model that will help you understand the events of mitosis. You can also use the model to demonstrate the effects of **nondisjunction** and **mutations.**

1. Write a definition for each boldface term in the paragraph above and for the following terms: chromatid, centromere, spindle fiber, cytokinesis.

   _____

   _____

   _____

   _____

2. Where in the human body do cells undergo mitosis?

   _____

3. How does a cell prepare to divide during interphase of the cell cycle?

   _____

   _____

| Modeling Mitosis *continued*

4. Based on the objectives for this lab, write a question you would like to explore about mitosis.

_____

_____

Procedure
PART A: DESIGN A MODEL

1. Work with the members of your lab group to design a model of a cell that uses the materials listed for this lab. Be sure your model cell has at least two pairs of chromosomes and is about to undergo mitosis.

> **You Choose**
> As you design your model, decide the following:
> **a.** what question you will explore
> **b.** how to construct a cell membrane
> **c.** how to show that your cell is diploid
> **d.** how to show the locations of at least two genes on each chromosome
> **e.** how to show that chromosomes are duplicated before mitosis begins

2. Write out the plan for building your model. Have your teacher approve the plan before you begin building the model.

_____

_____

_____

_____

3. Build the cell model your group designed. **CAUTION: Sharp or pointed objects can cause injury. Handle scissors carefully. Promptly notify your teacher of any injuries.** Use your model to demonstrate the phases of mitosis. Draw and label each phase you model.

Modeling Mitosis *continued*

4. Use your model to explore one of the questions written for step 4 of **Before You Begin.** Describe the steps you took to explore the question.

_____

_____

_____

_____

_____

PART B: TEST HYPOTHESES

Answer each of the following questions by writing a hypothesis. Use your model to test each hypothesis, and describe your results.

5. Cytokinesis follows mitosis. How will the size of each new cell that is formed following cytokinesis compare with that of the original cell?

_____

_____

6. Sometimes two chromatids fail to separate during mitosis. How might this failure affect the chromosome number of the two new cells?

_____

_____

_____

7. A mutation is a permanent change in a gene or chromosome. What effect might a mutation in a parent cell have on future generations of cells that result from the parent cell?

_____

_____

PART C: CLEANUP AND DISPOSAL

8. Dispose of paper and yarn scraps in the designated waste container.

9. Clean up your work area and all lab equipment. Return lab equipment to its proper place. Wash your hands thoroughly before you leave the lab and after you finish all work.

Modeling Mitosis *continued*

Analyze and Conclude

1. **Analyzing Results** How do the nuclei you made by modeling mitosis compare with the nucleus of the model cell you started with? Explain your result.

_____

_____

_____

_____

2. **Evaluating Methods** How could you modify your model to better illustrate the process of mitosis?

_____

_____

_____

_____

3. **Recognizing Patterns** How does the genetic makeup of the cells that result from mitosis compare with the genetic makeup of the original cell?

_____

_____

_____

_____

4. **Inferring Conclusions** How is mitosis important?

_____

_____

_____

5. **Further Inquiry** Write a new question about mitosis or the cell cycle that could be explored with your model.

_____

_____

_____

Exploration Lab

Modeling Meiosis

SKILLS
- Modeling
- Using scientific methods

OBJECTIVES
- **Describe** the events that occur in each stage of the process of meiosis.
- **Relate** the process of meiosis to genetic variation.

MATERIALS
- pipe cleaners of at least two different colors
- yarn
- wooden beads
- white labels
- scissors

Before You Begin

Meiosis is the process that results in the production of cells with half the normal number of chromosomes. It occurs in all organisms that undergo **sexual reproduction.** In this lab, you will build a model that will help you understand the events of meiosis. You can also use the model to demonstrate the effects of events such as **crossing-over** to explain results such as **genetic recombination.**

1. Write a definition for the following terms: homologous chromosomes, gamete.

_____

_____

2. In what organs in the human body do cells undergo meiosis?

_____

3. During interphase of the cell cycle, how does a cell prepare for dividing?

_____

_____

4. Based on the objectives for this lab, write a question you would like to explore about meiosis.

_____

_____

Modeling Meiosis *continued*

Procedure

PART A: DESIGN A MODEL

1. Work with the members of your lab group to design a model of a cell using the materials listed for this lab. Be sure that your model cell has at least two pairs of chromosomes.

2. Use a separate sheet of paper to write out the plan for building your model. Have your teacher approve the plan before you begin building the model.

3. Build the cell model your group designed. **CAUTION: Sharp or pointed objects can cause injury. Handle scissors carefully.** Use your model to demonstrate the phases of meiosis. Draw and label each phase you model.

4. Use your model to explore one of the questions written by your group for step 4 of **Before You Begin.** Describe the steps you took to explore your question.

You Choose

As you design your experiment, decide the following:

  a. what question you will explore

  b. how to construct a cell membrane

  c. how to show that your cell is diploid

  d. how to show the locations of at least two genes on each chromosome

  e. how to show that chromosomes are duplicated before meiosis begins

---

PART B: TEST HYPOTHESES

Answer each of the following questions by writing a hypothesis. Use your model to test each hypothesis, and describe your results.

5. In humans, gametes (eggs and sperm) result from meiosis. Will all gametes produced by one parent be identical?

_____

_____

_____

6. When an egg and a sperm fuse during sexual reproduction, the resulting cell (the first cell of a new organism) is called a zygote. How many copies of each chromosome and each gene will be found in a zygote?

_____

_____

_____

7. Crossing-over frequently occurs between the chromatids of homologous chromosomes during meiosis. Under what circumstances does crossing-over result in new combinations of genes in gametes?

_____

_____

_____

8. Synapsis (the pairing of homologous chromosomes) must occur before crossing-over can take place. How would the outcome of meiosis be different if synapsis did not occur?

_____

_____

_____

PART C: CLEANUP AND DISPOSAL

9. Dispose of paper and yarn scraps in the designated waste container.

10. Clean up your work area and all lab equipment. Return lab equipment to its proper place. Wash your hands thoroughly before you leave the lab and after finishing all work.

Analyze and Conclude

1. Analyzing Results How do the nuclei you made by modeling meiosis compare with the nucleus of the cell you started with? Explain your result.

_____

_____

2. Recognizing Relationships How are homologous chromosomes different from chromatids?

_____

_____

3. Forming Reasoned Opinions How is synapsis important to the outcome of meiosis? Explain.

_____

_____

Modeling Meiosis *continued*

4. Evaluating Methods How could you modify your model to better illustrate the process of meiosis?

_____

_____

_____

5. Drawing Conclusions How are the processes of meiosis similar to those of mitosis? How are they different?

_____

_____

_____

_____

_____

_____

6. Predicting Outcomes What would happen to the chromosome number of an organism's offspring if the gametes for sexual reproduction were made by mitosis instead of by meiosis?

_____

_____

7. Further Inquiry Write a new question about meiosis or sexual reproduction that could be explored with your model.

_____

_____

_____

Name _____ Class _____ Date _____

Modeling Monohybrid Crosses

MATERIALS

- lentils
- green peas
- 2 Petri dishes

SKILLS

- Predicting outcomes
- Calculating data
- Organizing data
- Analyzing data

OBJECTIVES

- **Predict** the genotypic and phenotypic ratios of offspring resulting from the random pairing of gametes.
- **Calculate** the genotypic ratio and phenotypic ratio among the offspring of a monohybrid cross.

Before You Begin

A **monohybrid cross** is a cross that involves one pair of contrasting traits. Different versions of a gene are called **alleles.** When two different alleles are present and one is expressed completely and the other is not, the expressed allele is **dominant** and the unexpressed allele is **recessive.**

1. Write a definition for each boldface term in the paragraph above. Use a separate sheet of paper.

2. Based on the objectives for this lab, write a question you would like to explore about heredity.

_____

_____

Procedure

PART A: SIMULATING A MONOHYBRID CROSS

1. You will model the random pairing of alleles by choosing lentils and peas from Petri dishes. These dried seeds will represent the alleles for seed color. A green pea will represent G, the dominant allele for green seeds, and a lentil will represent g, the recessive allele for yellow seeds.

Modeling Monohybrid Crosses *continued*

2. Each Petri dish will represent a parent. Label one Petri dish "female gametes" and the other Petri dish "male gametes." Place one green pea and one lentil in the Petri dish labeled "female gametes" and place one green pea and one lentil in the Petri dish labeled "male gametes."

3. Each parent contributes one allele to each offspring. Model a cross between these two parents by choosing a random pairing of the dried seeds from the two Petri dishes. Do this by simultaneously picking one seed from each Petri dish without looking. Place the pair of seeds together on the lab table. The pair of seeds represents the genotype of one offspring.

4. Record the genotype of the first offspring in Table A below.

Table A		
Gamete Pairings		
Trial	**Offspring genotype**	**Offspring phenotype**
1		
2		
3		
4		
5		
6		
7		
8		
9		
10		

5. Return the seeds to their original dishes and repeat step 3 nine more times. Record the genotype of each offspring in Table A.

6. Based on each offspring's genotype, determine and record each offspring's phenotype.

_____

_____

_____

PART B: CALCULATING GENOTYPIC AND PHENOTYPIC RATIOS

7. You will be using Table B to record your data.

8. Determine the genotypic and phenotypic ratios among the offspring. First count and record in Table B the number of homozygous dominant, heterozygous, and homozygous recessive individuals you recorded in Table A. Then

Modeling Monohybrid Crosses *continued*

record the number of offspring that produce green seeds and the number that produce yellow seeds under "Phenotypes" in Table B.

Table B		
Offspring Ratios		
Genotypes	**Total**	**Genotypic ratios**
Homozygous dominant *(GG)*		
Heterozygous *(Gg)*		_____ : _____ : _____
Homozygous recessive *(gg)*		
Phenotypes		**Phenotypic ratios**
Green seeds		
Yellow seeds		

9. Calculate the genotypic ratio for each genotype using the following equation:

$$\text{phenotypic ratio} = \frac{\text{number of offspring with a given genotype}}{\text{total number of offspring}}$$

10. Calculate the phenotypic ratio for each phenotype using the following equation:

$$\text{phenotypic ratio} = \frac{\text{number of offspring with a given phenotype}}{\text{total number of offspring}}$$

11. Now pool the data for the whole class, and record the data in Table C below.

12. Compare the class's sample with your small sample of 10. Calculate the genotypic and phenotypic ratios for the class data, and record them in Table C below.

13. Construct a Punnett square showing the parents and their offspring in your lab report.

14. Clean up your materials before leaving the lab.

Table C		
Offspring Ratios (Class Data)		
Genotypes	**Total**	**Genotypic ratios**
Homozygous dominant *(GG)*		
Heterozygous *(Gg)*		_____ : _____ : _____
Homozygous recessive *(gg)*		
Phenotypes		**Phenotypic ratios**
Green seeds		
Yellow seeds		

Modeling Monohybrid Crosses *continued*

Analyze and Conclude

1. **Summarizing Results** What character is being studied in this investigation?

_____

2. **Analyzing Data** What are the genotypes of the parents? Describe the genotypes of both parents using the terms *homozygous* or *heterozygous*, or both. Did Table B reflect a classic monohybrid-cross phenotypic ratio of 3:1?

_____

_____

_____

_____

_____

3. **Drawing Conclusions** If a genotypic ratio of 1:2:1 is observed, what must the genotypes of both parents be?

_____

4. **Predicting Patterns** Show what the genotypes of the parents would be if 50 percent of the offspring were green and 50 percent of the offspring were yellow.

_____

_____

5. **Further Inquiry** Construct a Punnett square for the cross of a heterozygous black guinea pig and an unknown guinea pig whose offspring include a recessive white-furred individual. Use a separate sheet of paper. What are the possible genotypes of the unknown parent?

Exploration Lab

Modeling DNA Structure

SKILLS
- Modeling
- Using scientific methods

OBJECTIVES
- **Design** and analyze a model of DNA.
- **Describe** how replication occurs.
- **Predict** the effect of errors during replication.

MATERIALS
- plastic soda straws, 3 cm sections
- metric ruler
- pushpins (red, blue, yellow, and green)
- paper clips

Before You Begin

DNA contains the instructions that cells need in order to make every **protein** required to carry out their activities and to survive. DNA is made of two strands of **nucleotides** twisted around each other in a **double helix.** The two strands are **complementary,** that is, the sequence of bases on one strand determines the sequence of bases on the other strand. The two strands are held together by hydrogen bonds.

In this lab, you will build a model to help you understand the structure of DNA. You can also use the DNA model to illustrate and explore processes such as **replication** and **mutation.**

1. Write a definition for each boldface term in the paragraphs above and for each of the following terms: replication fork, base-pairing rules. Use a separate sheet of paper.

2. Identify the three different components of a nucleotide.

_____

_____

3. Identify the four different nitrogen bases that can be found in DNA nucleotides.

_____

_____

4. Based on the objectives for this lab, write a question you would like to explore about DNA structure.

_____

_____

Procedure

PART A: DESIGN A MODEL

1. Work with the members of your lab group to design a model of DNA that uses the materials listed for this lab. Be sure that your model has at least 12 nucleotides on each strand.

> **You Choose**
> As you design your model, decide the following:
>
> **a.** what question you will explore
>
> **b.** how to use the straws, pushpins, and paper clips to represent the three components of a nucleotide
>
> **c.** how to link (bond) the nucleotides together
>
> **d.** in what order you will place the nucleotides on each strand

2. Write out the plan for building your model. Use a separate sheet of paper. Have your teacher approve the plan before you begin building the model.

3. ◆ Build the DNA model your group designed. **CAUTION: Sharp or pointed objects may cause injury. Handle pushpins carefully.** Sketch and label the parts of your DNA model.

4. Use your model to explore one of the questions written for step 4 of **Before You Begin.**

PART B: DNA REPLICATION

5. Discuss with your lab group how the model you built for Part A may be used to illustrate the process of replication.

6. Write a question you would like to explore about replication. Use your model to explore the question you wrote. On a separate sheet of paper, sketch and label the steps of replication.

_____

_____

| Modeling DNA Structure *continued*

PART C: TEST HYPOTHESIS

Answer each of the following questions by writing a hypothesis. Use your model to test each hypothesis, and describe your results.

7. Mitosis follows replication. How might the cells produced by mitosis be affected if nucleotides on one DNA strand were incorrectly paired during replication?

_____

_____

8. What would happen if only one strand in a DNA molecule were copied during replication?

_____

_____

PART D: CLEANUP AND DISPOSAL

9. Dispose of damaged pushpins in the designated waste container.

10. Clean up your work area and all lab equipment. Return lab equipment to its proper place. Wash your hands thoroughly before you leave the lab and after you finish all work.

Analyze and Conclude

1. Analyzing Results In your original DNA model, were the two strands identical to each other?

_____

2. Relating Concepts How does DNA structure ensure that the two DNA molecules made by replication are the same as the original DNA molecule?

_____

_____

_____

3. Drawing Conclusions Did the two DNA molecules you made in step 6 have the same nitrogen-base sequence as your original model DNA molecule?

_____

_____

4. Inferring Relationships The order of nitrogen bases on a DNA strand is a code for making proteins. What does this mean has happened to the "code" in one of the DNA molecules you made in step 7?

_____

_____

5. Predicting Outcomes What would happen if the DNA in a cell that is about to divide were not replicated?

_____

_____

_____

6. Inferring Information What are the advantages of having DNA remain in the nucleus of a cell?

_____

_____

_____

7. Further Inquiry Write a new question about DNA that could be explored with your model.

_____

_____

_____

Exploration Lab

Modeling Protein Synthesis

SKILLS

- Modeling
- Using scientific methods

OBJECTIVES

- **Compare** and **Contrast** the structure and function of DNA and RNA.
- **Model** protein synthesis.
- **Demonstrate** how a mutation can affect a protein.

MATERIALS

- masking tape
- plastic soda-straw pieces of one color
- plastic soda-straw pieces of a different color
- paper clips
- pushpins of five different colors
- marking pens of the same colors as the pushpins
- 3×5 in. note cards
- oval-shaped card
- transparent tape

Before You Begin

The nature of a **protein** is determined by the sequence of amino acids in its structure. During **protein synthesis**, the sequence of nitrogen bases in an **mRNA** molecule is used to assemble **amino acids** into a protein chain.

A **mutation** is a change in the nitrogen-base sequence of DNA. Many mutations lead to altered or defective proteins. For example, the genetic blood disorder **sickle cell anemia** is caused by a mutation in the gene for **hemoglobin**.

In this lab, you will build models that will help you understand how protein synthesis occurs. You can also use the models to explore how a mutation affects a protein.

1. Write a definition for each boldface term in the paragraph above and for each of the following terms: transcription, translation, tRNA, ribosome, codon, anticodon. Use a separate sheet of paper.

2. Describe three differences between DNA and RNA.

_____

_____

_____

_____

3. Based on the objectives for this lab, write a question you would like to explore about protein synthesis.

_____

_____

Procedure

PART A: DESIGN A MODEL

1. Work with the members of your lab group to design models of DNA, RNA, and a cell. Use the materials listed for this lab.

> **You Choose**
>
> As you design your models, decide the following:
>
> **a.** what question you will explore
>
> **b.** how to represent DNA nucleotides
>
> **c.** how to represent RNA nucleotides
>
> **d.** how to represent five different nitrogen bases
>
> **e.** how to link (bond) nucleotides together
>
> **f.** how to represent tRNA molecules with amino acids
>
> **g.** how to represent the locations of DNA and ribosomes

2. Write out the plan for building your models. Have your teacher approve the plan before you begin building the models.

_____

_____

_____

_____

Modeling Protein Synthesis *continued*

3. Build the models your group designed. **CAUTION: Sharp or pointed objects may cause injury. Handle pushpins carefully.** Start your model of DNA with a strand of nucleotides that has the following sequence of nitrogen bases: TTTGGTCTCCTC.

PART B: MODEL PROTEIN SYNTHESIS

4. Use your models to demonstrate how transcription and translation occur. Draw and label the steps of each process on a separate sheet of paper.

5. Use your models to explore one of the questions written for step 3 of **Before You Begin.**

_____

_____

_____

_____

PART C: TEST HYPOTHESIS

Answer each of the following questions by writing a hypothesis. Use your models to test each hypothesis, and describe your results.

6. The DNA model you built for step 3 represents a portion of a gene for hemoglobin. Sickle cell anemia results from the substitution of an A for the T in the third codon of the nitrogen-base sequence given in step 3. How will this substitution affect a hemoglobin molecule?

_____

_____

7. The addition of a nucleotide to a strand of DNA is a type of mutation called an *insertion*. What happens when an insertion occurs in the first codon in a DNA strand, before the DNA strand is transcribed?

_____

_____

PART D: CLEANUP AND DISPOSAL

8. Dispose of damaged pushpins in the designated waste container.

9. Clean up your work area and all lab equipment. Return lab equipment to its proper place. Wash your hands thoroughly before you leave the lab and after you finish all work.

Modeling Protein Synthesis *continued*

Analyze and Conclude

1. **Comparing Structures** How did the nitrogen-base sequence of the mRNA you made compare with that of the DNA it was transcribed from?

_____

_____

2. **Recognizing Relationships** How is the nitrogen-base sequence of a gene related to the structure of a protein?

_____

_____

3. **Recognizing Patterns** What is the relationship between the anticodon of a tRNA and the amino acid the tRNA carries?

_____

_____

4. **Drawing Conclusions** How does a mutation in the gene for a protein affect the protein?

_____

_____

_____

5. **Further Inquiry** Write a new question about protein synthesis that could be explored with your model.

_____

_____

_____

Exploration Lab

Modeling Recombinant DNA

SKILLS

- Modeling
- Comparing

OBJECTIVES

- **Construct** a model that can be used to explore the process of genetic engineering.
- **Describe** how recombinant DNA is made.

MATERIALS

- paper clips (56)
- plastic soda straw pieces (56)
- pushpins (15 red, 15 green, 13 blue, and 13 yellow)

Before You Begin

Genetic engineering is the process of taking a gene from one organism and inserting it into the DNA of another organism. The gene is delivered by a **vector**, such as a virus, or a bacterial **plasmid.**

First, a fragment of a chromosome that contains the gene is isolated by using a **restriction enzyme**, which cuts DNA at a specific nucleotide-base sequence. Some restriction enzymes cut DNA unevenly, producing single-stranded **sticky ends.** The DNA of the vector is cut by the same restriction enzyme. Next, the chromosome fragment is mixed with the cut DNA of the vector. Finally, an enzyme called **DNA ligase** joins the ends of the two types of cut DNA, producing **recombinant DNA.**

In this lab, you will model genetic engineering techniques. You will simulate the making of recombinant DNA that has a human gene inserted into the DNA of a plasmid.

1. Write a definition for each boldface term in the paragraph above and for the term *base-pairing rules*. Use a separate sheet of paper.

2. Based on the objectives for this lab, write a question you would like to explore about the process of genetic engineering.

_____

_____

Procedure
PART A: MODEL GENETIC ENGINEERING

1. Make 56 model nucleotides. To make a nucleotide, insert a pushpin mid-way along the length of a 3 cm piece of a soda straw. **CAUTION: Handle pushpins carefully. Pointed objects can cause injury.** Push a paper clip into one end of the soda-straw piece until it touches the pushpin.

2. Begin a model of a bacterial plasmid by arranging nucleotides for one DNA strand in the following order: blue, red, green, yellow, red, red, blue, blue, green, red, blue, green, red, blue, blue, green, yellow, and red. Join two adjacent nucleotides by inserting the paper clip end of one into the open end of the other.

3. Using your first DNA strand and the base-pairing rules, build the complementary strand of plasmid DNA. **Note:** *Yellow is complementary to blue, and green is complementary to red.*

4. Complete your model of a circular plasmid by joining the opposite ends of each DNA strand. Make a sketch showing the sequence of bases in your model plasmid. Use the abbreviations B, Y, G, and R for the pushpin colors.

5. Begin a model of a human chromosome fragment made by a restriction enzyme. Place nucleotides for one DNA strand in the following order: BBRRYGGBRY. Build the second DNA strand by arranging the remaining nucleotides in the following order: BRRYGBYYGG.

6. Match the complementary portions of the two strands of DNA you made in step 5. Pair as many base pairs in a row as you can. Make a sketch showing the sequence of bases in your model of a human chromosome fragment.

Modeling Recombinant DNA *continued*

7. Imagine that the restriction enzyme that cut the human chromosome fragment you made in steps 5 and 6 is moving around your model plasmid until it finds the sequence YRRBBG and its complementary sequence, BGGYYR. This restriction enzyme cuts each sequence between a B and a G. Find such a section in your sketch of your model plasmid's DNA.

8. Simulate the action of the restriction enzyme on the section you identified in step 7. Open both strands of your model plasmid's DNA by pulling apart the adjacent green and blue nucleotides in each strand. Make a sketch of the split plasmid DNA molecule.

9. Move your model human DNA fragment into the break in your model plasmid's DNA molecule. Imagine that a ligase joins the ends of the human and plasmid DNA. Make a sketch of your final model DNA molecule.

PART B: CLEANUP AND DISPOSAL

10. Dispose of damaged pushpins in the designated waste container.

11. Clean up your work area and all lab equipment. Return lab equipment to its proper place. Wash your hands thoroughly before you leave the lab and after you finish all work.

Modeling Recombinant DNA *continued*

Analyze and Conclude

1. Comparing Structures Compare your models of plasmid DNA and human DNA.

_____

_____

_____

_____

_____

_____

2. Relating Concepts What do the sections of four unpaired nucleotides in your model human DNA fragment represent?

_____

_____

3. Comparing Structures How did your original model plasmid DNA molecule differ from your final model DNA molecule?

_____

_____

_____

4. Drawing Conclusions What does the molecule you made in step 9 represent?

_____

_____

5. Further Inquiry Write a new question that could be explored with another investigation.

_____

_____

Name _____ Class _____ Date _____

Making a Timeline of Life on Earth

SKILLS

- Observing
- Inferring relationships
- Organizing data

OBJECTIVES

- **Compare** and **contrast** the distinguishing characteristics of representative organisms of the six kingdoms.
- **Organize** the appearance of life on Earth in a timeline.

MATERIALS

- adding-machine tape (5 m roll)
- meterstick
- colored pens or pencils
- photographs or drawings of organisms from ancient Earth to present day

Before You Begin

About 4.5 billion years ago, Earth was a ball of molten rock. As the surface cooled, a rocky crust formed and water vapor in the atmosphere condensed to form rain. By 3.9 billion years ago, oceans covered much of Earth's surface. Rocks formed in these oceans contain **fossils** of bacterial cells that lived about 3.5 billion years ago. The **fossil record** shows a progression of life-forms and contains evidence of many changes in Earth's surface and atmosphere.

In this lab, you will make a **timeline** showing the major events in Earth's history and in the history of life on Earth, such as the evolution of new groups of organisms and the mass extinctions. This timeline can be used to study how living things have changed over time.

1. Write a definition for each boldface term in the paragraphs above. Use a separate sheet of paper.

2. Record your data in the data table provided.

3. Based on the objectives for this lab, write a question you would like to explore about the history of life on Earth.

_____

_____

Making a Timeline of Life on Earth *continued*

Procedure

PART A: MAKING A TIMELINE

1. Make a mark every 20 cm along a 5 m length of adding-machine tape. Label one end of the tape "5 billion years ago" and the other end "Today." Write "20 cm = 200 million years" near the beginning of your timeline.

2. Locate and label a point representing the origin of Earth on your timeline. Use your textbook as a reference. Also locate and label the periods of the geologic time scale.

3. Using your textbook as a reference, mark the following events on your timeline: the first cyanobacteria appear; oxygen enters the atmosphere; the five mass extinctions; the first eukaryotes appear; the first multicellular organisms appear; the first vertebrates appear; the first plants, fungi, and land animals appear; the first dinosaurs and mammals appear; the first flowering plants appear; the first humans appear.

4. Look at the photographs of organisms provided by your teacher. Identify the major characteristics of each organism. Record your observations in the data table below.

Data Table		
Organism	**Kingdom**	**Characteristics/adaptation for life on Earth**

Making a Timeline of Life on Earth *continued*

5. Lay out your timeline on the floor in your classroom. Place photographs (or drawings) of the organisms you examined on your timeline to show when they appeared on Earth.

6. Fold the timeline at the mark representing 4.8 billion years ago. This leaves 24 segments, each representing 200 million years on your timeline. Now you can think of each segment as 1 hour in a 24-hour day.

7. When you are finished, walk slowly along your timeline. Note the sequence of events in the history of life on Earth and the relative amount of time between each event.

PART B: CLEANUP AND DISPOSAL

8. Dispose of paper scraps in the designated waste container.

9. Clean up your work area and all lab equipment. Return lab equipment to its proper place.

ANALYZE AND CONCLUDE

1. **Analyzing Information** Think of each segment of your timeline as 1 hour in a 24-hour day as you answer each of the following questions.

   a. How long has life existed on Earth?

   _____

   b. For what part of the day did only unicellular life-forms exist?

   _____

   c. At what time of day did the first plants appear on Earth?

   _____

   d. At what time of day did mammals appear on Earth?

   _____

2. **Summarizing Information** Identify the major developments in life-forms that have occurred over the last 3.5 billion years.

   _____

   _____

   _____

   _____

Making a Timeline of Life on Earth *continued*

3. Inferring Relationships How do mass extinctions appear to be related to the appearance of new major groups of organisms?

_____

_____

4. Justifying Conclusions Cyanobacteria are thought to be responsible for adding oxygen to Earth's atmosphere. Use your timeline to justify this conclusion.

_____

_____

_____

5. Calculating Determine the amount of time, as a percentage of the time that life has existed on Earth, that humans (Homo sapiens) have existed.

6. Further Inquiry Write a new question about the history of life on Earth that could be explored in another investigation.

_____

_____

_____

Exploration Lab

Modeling Natural Selection

SKILLS

• Modeling a process
• Inferring relationships

OBJECTIVES

• **Model** the process of selection.
• **Relate** favorable mutations to selection and evolution.

MATERIALS

• scissors
• construction paper
• cellophane tape
• soda straws
• felt-tip marker
• meterstick or tape measure
• penny or other coin
• six-sided die

Before You Begin

Natural selection occurs when organisms that have certain **traits** survive to reproduce more than organisms that lack those traits do. A population evolves when individuals with different **genotypes** survive or reproduce at different rates. In this lab, you will model the selection of favorable traits in a new generation by using a paper model of a bird—the fictitious Egyptian origami bird (*Avis papyrus*), which lives in dry regions of North Africa. Assume that only birds that can successfully fly the long distances between water sources will live long enough to breed successfully.

1. Write a definition for each boldface term in the preceding paragraph. Use a separate sheet of paper.

2. You will be using the data table provided to record your data.

3. Based on the objectives for this lab, write a question you would like to explore about the process of selection.

_____

_____

Modeling Natural Selection *continued*

Procedure
PART A: PARENTAL GENERATION

1. Cut two strips of paper, 2 × 20 cm each. Make a loop with one strip of paper, letting the paper overlap by 1 cm, and tape the loop closed. Repeat for the other strip.

2. Tape one loop 3 cm from each end of the straw, as shown. Mark the front end of the bird with a felt-tip marker. This bird represents the parental generation.

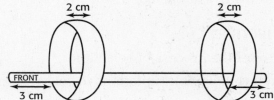

3. Test how far your parent bird can fly by releasing it with a gentle overhand pitch. Test the bird twice. Record the bird's average flight distance in the data table on the next page.

PART B: FIRST (F₁) GENERATION

4. Each origami bird lays a clutch of three eggs. Assume that one of the chicks is a clone of the parent. Use the parent to represent this chick in step 6.

5. Make two more chicks. Assume that these chicks have mutations. Follow Steps A–C below for each chick to determine the effects of its mutation.

    Step A Flip a coin to determine which end is affected by a mutation.

    Heads = anterior (front)

    Tails = posterior (back)

    Step B Throw a die to determine how the mutation affects the wing.

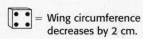 = Wing position moves 1 cm toward the end of the straw.   = Wing circumference decreases by 2 cm.

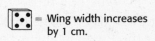 = Wing position moves 1 cm toward the middle of the straw.   = Wing width increases by 1 cm.

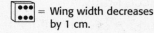 = Wing circumference increases by 2 cm.   = Wing width decreases by 1 cm.

    Step C A mutation is lethal if it causes a wing to fall off the straw or a wing with a circumference smaller than that of the straw. If you get a lethal mutation, disregard it and produce another chick.

6. Record the mutations and the wing dimensions of each offspring.

7. Test each bird twice by releasing it with a gentle overhand pitch. Release the birds as uniformly as possible. Record the distance each bird flies. The most successful bird is the one that flies the farthest.

Name _____ Class _____ Date _____

Modeling Natural Selection *continued*

			Anterior wing (cm)			Posterior wing (cm)			
Bird	**Coin flip (H or T)**	**Die throw (1–6)**	**Width**	**Circum.**	**Distance from front**	**Width**	**Circum.**	**Distance from back**	**Average distance flown (m)**
Parent	NA	NA	2	19	3	2	19	3	
Generation 1									
Chick 1									
Chick 2									
Chick 3									
Generation 2									
Chick 1									
Chick 2									
Chick 3									
Generation 3									
Chick 1									
Chick 2									
Chick 3									
Generation 4									
Chick 1									
Chick 2									
Chick 3									
Generation 5									
Chick 1									
Chick 2									
Chick 3									
Generation 6									
Chick 1									
Chick 2									
Chick 3									
Generation 7									
Chick 1									
Chick 2									
Chick 3									
Generation 8									
Chick 1									
Chick 2									
Chick 3									
Generation 9									
Chick 1									
Chick 2									
Chick 3									

Data Table

Modeling Natural Selection *continued*

PART C: SUBSEQUENT GENERATIONS

8. Assume that the most successful bird in the previous generation is the sole parent of the next generation. Repeat steps 4–7 using this bird.

9. Continue to breed, test, and record data for eight more generations.

PART D: CLEANUP AND DISPOSAL

10. Dispose of paper scraps in the designated waste container.

11. Clean up your work area and all lab equipment. Return lab equipment to its proper place. Wash your hands thoroughly before you leave the lab and after you finish all work.

Analyze and Conclude

1. Analyzing Results Did the birds you made by modeling natural selection fly farther than the first bird you made?

_____

_____

2. Inferring Conclusions How might this lab help explain the variety of species of Galápagos finches?

_____

_____

_____

_____

3. Further Inquiry Write another question about natural selection that could be explored with another investigation.

_____

_____

_____

_____

Skills Practice Lab

Making a Dichotomous Key

SKILLS

- Identifying and comparing
- Organizing data

OBJECTIVES

- **Identify** objects using dichotomous keys.
- **Design** a dichotomous key for a group of objects.

MATERIALS

- 6 to 10 objects found in the classroom
  (e.g., shoes, books, writing instruments)
- stick-on labels
- pencil

Before You Begin

One way to identify an unknown organism is to use an **identification key,** which contains the major characteristics of groups of organisms. A **dichotomous key** is an identification key that contains pairs of contrasting descriptions. After each description, a key either directs the user to another pair of descriptions or identifies an object. In this lab, you will design and use a dichotomous key. A dichotomous key can be written for any group of objects.

1. Write a definition for each boldface term in the paragraph above.

_____

_____

_____

_____

2. Based on the objectives for this lab, write a question you would like to explore about making or using a dichotomous key.

_____

_____

Procedure

PART A: USING A DICHOTOMOUS KEY

1. Use the **Key to Forest Trees** to identify the tree that produced each of the leaves shown. Identify one leaf at a time. Always start with the first pair of statements (**1***a* and **1***b*). Follow the direction beside the statement that describes the leaf. Proceed through the key until you get to the name of a tree.

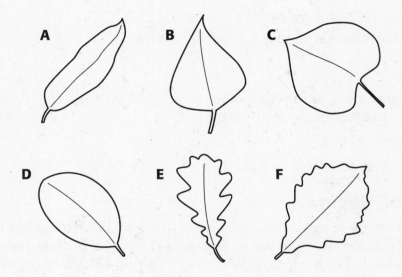

Key to Forest Trees	
1*a* Leaf edge has no teeth, waves, or lobes	go to **2**
1*b* Leaf edge has teeth, waves, or lobes	go to **3**
2*a* Leaf has a bristle at its tip	**shingle oak**
2*b* Leaf has no bristle at its tip	go to **4**
3*a* Leaf edge is toothed	**Lombardy poplar**
3*b* Leaf edge has waves or lobes	go to **5**
4*a* Leaf is heart-shaped	**red bud**
4*b* Leaf is not heart-shaped	**live oak**
5*a* Leaf edge has lobes	**English oak**
5*b* Leaf edge has waves	**chestnut oak**

PART B: DESIGN A DICHOTOMOUS KEY

2. Work with the members of your lab group to design a dichotomous key using the materials listed for this lab.

> **You Choose**
>
> As you design your key, decide the following:
>
> **a.** what question you will explore
>
> **b.** what objects your key will identify
>
> **c.** how you will label personal property
>
> **d.** what distinguishing characteristics the objects have
>
> **e.** which characteristics to use in your key
>
> **f.** how you will organize the data you will need for writing your key

3. Before you begin writing your key, have your teacher approve the objects your group has decided to work with.

4. Using the **Key to Forest Trees** as a guide, write a key for the objects your group selected. Remember, a dichotomous key includes pairs of contrasting descriptions.

5. Use your key to explore one of the questions written for step 2 of **Before You Begin.**

_____

_____

6. After each group has completed step 5, exchange keys and the objects they identify with another group. Use the key you receive to identify the objects. If the key does not work, return it to the group so corrections can be made.

PART C: CLEANUP

7. Clean up your work area and all lab equipment. Return lab equipment to its proper place. Wash your hands thoroughly before you leave the lab and after you finish all work.

Analyze and Conclude

1. Drawing Conclusions What tree produced each of the leaves shown in this lab?

_____

_____

_____

_____

2. Forming Hypotheses What other characteristics might be used to identify leaves using a dichotomous key?

_____

_____

_____

_____

3. Analyzing Methods How was the key your group designed dichotomous?

_____

_____

_____

4. Evaluating Results Were you able to use another group's key to identify the objects for which it was written? If not, describe the problems you encountered.

_____

_____

_____

_____

5. Analyzing Methods Does a dichotomous key begin with general descriptions and then proceed to more specific descriptions or vice versa? Explain your answer, giving an example from your key.

_____

_____

_____

6. Further Inquiry Write a new question about making or using keys that could be explored with another investigation.

_____

_____

_____

Observing How Natural Selection Affects a Population

SKILLS
- Using scientific methods
- Collecting, graphing, and analyzing data

OBJECTIVES
- **Measure** and collect data for a trait in a population.
- **Graph** a frequency distribution curve of your data.
- **Analyze** your data by determining its mean, median, mode, and range.
- **Predict** how natural selection can affect the variation in a population.

MATERIALS
- metric ruler
- graph paper (optional)
- green beans or snow peas
- calculator
- balance

Before You Begin

Natural selection can occur when there is **variation** in a **population**. You can analyze the variation in certain traits of a population by determining the mean, median, mode, and range of the data collected on several individuals. The **mean** is the sum of all data values divided by the number of values. The **median** is the midpoint in a series of values. The **mode** is the most frequently occurring value. The **range** is the difference between the largest and smallest values. The variation in a characteristic can be visualized with a **frequency distribution curve**. Two kinds of natural selection—**stabilizing selection** and **directional selection**—can influence the frequency and distribution of traits in a population. This changes the shape of a frequency distribution curve. In this lab, you will investigate variation in fruits and seeds.

1. Write a definition for each boldface term in the paragraph above. Use a separate sheet of paper.

2. Based on the objectives for this lab, write a question you would like to explore about variation in green beans or snow peas.

_____

_____

| Observing How Natural Selection Affects a Population *continued*

Procedure
PART A: DESIGN AN EXPERIMENT

1. Work with the members of your lab group to explore one of the questions written for step 2 of **Before You Begin.** To explore the question, design an experiment that uses the materials listed for this lab.

> **You Choose**
>
> As you design your experiment, decide the following:
>
> **a.** what question you will explore
>
> **b.** what hypothesis you will test
>
> **c.** which trait (length, color, weight, etc.) you will measure
>
> **d.** how you will measure the trait
>
> **e.** how many members of the population you will measure (keep in mind that the more data you gather, the more revealing your frequency distribution curve will be)
>
> **f.** what data you will record in your data table

2. Write a procedure for your experiment. Make a list of all the safety precautions you will take. Have your teacher approve your procedure and safety precautions before you begin the experiment.

_____

_____

_____

_____

_____

_____

3. Conduct your experiment.

PART B: CLEANUP AND DISPOSAL

4. Dispose of seeds in the designated waste containers. Do not put lab materials in the trash unless your teacher tells you to do so.

5. Clean up your work area and all lab equipment. Return lab equipment to its proper place. Wash your hands thoroughly before you leave the lab and after you finish all work.

Observing How Natural Selection Affects a Population *continued*

Analyze and Conclude

1. **Summarizing Results** Make a frequency distribution curve of your data. Plot the trait you measured on the x-axis (horizontal axis) and the number of times that trait occurred in your population on the y-axis (vertical axis).

2. **Calculating** Determine the mean, median, mode, and range of the data for the trait you studied.

_____

_____

_____

_____

3. **Analyzing Results** How does the mean differ from the mode in your population?

_____

_____

_____

4. **Drawing Conclusions** What type of selection appears to have produced the type of variation observed in your experiment?

_____

_____

_____

_____

Observing How Natural Selection Affects a Population *continued*

5. **Evaluating Data** The graph below shows the distribution of wing length in a population of birds on an island. Notice that the mean and the mode are quite different. Is the mean always useful in describing traits in a population? Explain.

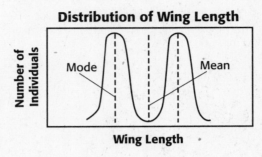

Distribution of Wing Length

_____

_____

_____

_____

_____

6. **Forming Hypotheses** What type of selection (stabilizing or directional) would be indicated if the mean of a trait you measured shifted, over time, to the right of a frequency distribution graph?

_____

7. **Further Inquiry** Write a new question about variation in populations that could be explored in another investigation.

_____

_____

_____

_____

Skills Practice Lab

Surveying Kingdom Diversity

SKILLS

• Using a microscope

• Comparing

OBJECTIVES

• **Observe** representatives of each of the six kingdoms.

• **Compare** and **contrast** the organisms within a kingdom.

• **Analyze** the similarities and differences among the six kingdoms.

MATERIALS

• specimens from each of the six kingdoms

• compound microscopes

• hand lenses or stereomicroscopes

SAFETY

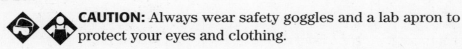 **CAUTION:** Always wear safety goggles and a lab apron to protect your eyes and clothing.

CAUTION: Do not touch or taste any chemicals. Know the location of the emergency shower and eyewash station and how to use them. If you get a chemical on your skin or clothing, wash it off at the sink while calling to the teacher. Notify the teacher of a spill. Spills should be cleaned up promptly, according to your teacher's directions.

CAUTION: Glassware is fragile. Notify the teacher of broken glass or cuts. Do not clean up broken glass or spills with broken glass unless the teacher tells you to do so.

Before You Begin

Many biologists classify living things into six **kingdoms.** The organisms in a kingdom have fundamental characteristics in common. For example, the organisms of two kingdoms are made of **prokaryotic cells,** while the organisms in the other four kingdoms are made of **eukaryotic cells.** Some kingdoms contain only **unicellular** or **colonial** organisms, while others contain only **multicellular** organisms. In this lab, you will examine representatives of six kingdoms of organisms. You will see that each kingdom is distinct from the others.

1. Write a definition for each boldface term in the previous paragraph and for each of the following terms: tissue, organ, organ system, autotroph, heterotroph. Use a separate sheet of paper.

2. You will be using the data table provided to record your data.

3. Based on the objectives for this lab, write a question you would like to explore about the kingdoms of organisms.

_____

_____

Procedure

PART A: CONDUCTING A SURVEY

1. Put on safety goggles and a lab apron.

2. Visit the station for each kingdom and examine the specimens there. Answer the questions, and record observations in your data table.

Data Table				
Kingdom name	**Type of cells**	**Level of organization**	**Other characteristics**	**Examples**

3. Archaebacteria Examine the prepared slides.

a. What does a microscope reveal about the structure of archaebacteria?

_____

b. How do these organisms get energy for life processes?

_____

4. Eubacteria Examine the prepared slides.

a. What does a microscope reveal about the structure of eubacteria?

_____

b. Would you consider *Anabaena* to be unicellular or multicellular? Explain.

_____

_____

Surveying Kingdom Diversity *continued*

c. How does *Anabaena* appear to obtain energy for life processes? Explain.

_____

_____

5. Protists Examine the prepared slides.

a. What does a microscope reveal about the structure of protozoans?

_____

_____

_____

b. How do protozoans appear to obtain energy for life processes? Explain.

_____

_____

c. Are the algae unicellular or multicellular? Explain.

_____

_____

d. How do algae differ from protozoans?

_____

_____

6. Fungi Examine the specimens.

a. Are fungi unicellular or multicellular? Explain.

_____

_____

b. What does a microscope reveal about the structure of fungi?

_____

_____

c. How do the fungi appear to obtain energy for life processes? Explain.

_____

_____

7. Plants Examine the specimens.

a. What is the most striking characteristic shared by these plants?

_____

_____

b. What does a microscope reveal about the structure of plants?

_____

_____

8. Animals Examine the specimens.

a. What is the most striking characteristic shared by these animals?

_____

_____

b. What is the most striking difference among these animals?

_____

_____

PART B: CLEANUP AND DISPOSAL

9. Dispose of broken glass and solutions in the designated waste containers. Do not pour chemicals down the drain or put lab materials in the trash unless your teacher tells you to do so.

10. Wash your hands thoroughly before you leave the lab and after you finish all work.

Analyze and Conclude

1. Summarizing Data What are the main differences observed among the six kingdoms?

_____

_____

_____

_____

2. Recognizing Patterns How does the size of prokaryotic cells compare with the cell size in the other kingdoms?

_____

_____

Surveying Kingdom Diversity *continued*

3. Analyzing Methods How did you determine the cell type for each kingdom?

_____

_____

4. Inferring Conclusions Which kingdom exhibits the most diversity?

_____

_____

5. Further Inquiry Write a new question about the kingdoms of life that could be explored with another investigation.

_____

_____

Exploration Lab

Modeling Ecosystem Change over Time

SKILLS

• Using scientific methods
• Modeling
• Observing

OBJECTIVES

• **Construct** a model ecosystem.
• **Observe** the interactions of organisms in a model ecosystem.
• **Predict** how the number of each species in a model ecosystem will change over time.
• **Compare** a model ecosystem with a natural ecosystem.

MATERIALS

• coarse sand or pea gravel
• large glass jar with a lid or terrarium
• soil
• pinch of grass seeds
• pinch of clover seeds
• rolled oats
• mung bean seeds
• earthworms
• isopods (pill bugs)
• mealworms (beetle larva)
• crickets

Before You Begin

Organisms in an **ecosystem** interact with each other and with their environment. One of the interactions that occurs among the organisms in an ecosystem is feeding. A **food web** describes the feeding relationships among the organisms in an ecosystem. In this lab, you will model a natural ecosystem by building **a closed ecosystem** in a bottle or a jar. You will then observe the interactions of the organisms in the ecosystem and note any changes that occur over time.

  1. Write a definition for each boldface term in the paragraph above and for each of the following terms: producer, decomposer, consumer, herbivore, carnivore, trophic level. Use a separate sheet of paper.

Modeling Ecosystem Change over Time *continued*

2. Based on the objectives for this lab, write a question you would like to explore about ecosystems.

_____

_____

Procedure

PART A: BUILDING AN ECOSYSTEM IN A JAR

1. Place 2 in. of sand or pea gravel in the bottom of a large, clean glass jar with a lid. **CAUTION: Glassware is fragile. Notify your teacher promptly of any broken glass or cuts. Do not clean up broken glass or spills with broken glass unless your teacher tells you to do so.** Cover the gravel with 2 in. of soil.

2. Sprinkle the seeds of two or three types of small plants, such as grasses and clovers, on the surface of the soil. Put a lid on the jar, and place it in indirect sunlight. Let the jar remain undisturbed for a week.

3. After one week, place a handful of rolled oats in the jar. Place the mealworms in the oats, and then place the other animals into the jar and replace the lid. Place the lid on the jar loosely to enable air entry.

You Choose

As you design your experiment, decide the following:

    a. what question you will explore

    b. what hypothesis you will test

    c. how you will plant the seeds

    d. where you will place the ecosystem for one week so that it remains undisturbed and in indirect sunlight

    e. how often you will add water to the ecosystem after the first week

    f. how many of each organism you will use

    g. what data you will record in your data table

PART B: DESIGN AN EXPERIMENT

4. Work with the members of your lab group to explore one of the questions written for step 2 of **Before You Begin.** To explore the question, design an experiment that uses the materials listed for this lab.

5. Write a procedure for your experiment. Make a list of all the safety precautions you will take. Have your teacher approve your procedure and safety precautions before you begin the experiment.

6. Set up your group's experiment. Conduct your experiment for at least 14 days.

Modeling Ecosystem Change over Time *continued*

PART C: CLEANUP AND DISPOSAL

7. Dispose of solutions, broken glass, and other materials in the designated waste containers. Do not put lab materials in the trash unless your teacher tells you to do so.

8. Clean up your work area and all lab equipment. Return lab equipment to its proper place. Wash your hands thoroughly before you leave the lab and after you finish all work.

Analyze and Conclude

1. **Summarizing Results** Make graphs showing how the number of individuals of each species in your ecosystem changed over time. Plot time on the x-axis and the number of organisms on the y-axis.

_____

_____

_____

2. **Analyzing Results** How did your results compare with your hypothesis? Explain any differences.

_____

_____

_____

3. **Inferring Conclusions** Construct a food web for the ecosystem you observed.

_____

_____

_____

_____

4. **Recognizing Relationships** Does your model ecosystem resemble a natural ecosystem? Explain.

_____

_____

_____

_____

_____

Modeling Ecosystem Change over Time *continued*

5. Analyzing Methods How might you have built your model ecosystem differently to better represent a natural ecosystem?

_____

_____

_____

_____

6. Evaluating Methods Was your model ecosystem truly a "closed ecosystem"? List your model's strengths and weaknesses as a closed ecosystem.

_____

_____

_____

_____

_____

_____

7. Further Inquiry Write a new question about ecosystems that you could explore with another investigation.

_____

_____

_____

Observing How Brine Shrimp Select a Habitat

SKILLS

- Using scientific methods
- Collecting, organizing, and graphing data

OBJECTIVES

- **Observe** the behavior of brine shrimp.
- **Assess** the effect of environmental variables on habitat selection by brine shrimp.

MATERIALS

- clear, flexible plastic tubing
- metric ruler
- marking pen
- corks to fit tubing
- brine shrimp culture
- screw clamps
- test tubes with stoppers and test-tube rack
- pipet
- Petri dish

- Detain™ or methyl cellulose
- aluminum foil
- calculator
- fluorescent lamp or grow light
- funnel
- graduated cylinder or beaker
- hot-water bag
- ice bag
- pieces of screen
- tape

Before You Begin

Different organisms are adapted for life in different **habitats.** For example, **brine shrimp** are small crustaceans that live in salt lakes. Given a choice, organisms select habitats that provide the conditions (e.g., temperature, light, pH, salinity) to which they are adapted. In this lab, you will investigate habitat selection by brine shrimp and determine which environmental conditions they prefer.

1. Write a definition for each boldface term in the paragraph above.

_____

_____

_____

2. Based on the objectives for this lab, write a question you would like to explore about habitat selection by brine shrimp.

_____

_____

Observing How Brine Shrimp Select a Habitat *continued*

Procedure
PART A: MAKING AND SAMPLING A TEST CHAMBER

1. Divide a piece of plastic tubing into 4 sections by making a mark at 12 cm, 22 cm, and 32 cm from one end. Label the sections *1, 2, 3,* and *4.*

2. Place a cork in one end of the tubing. Then transfer 50 mL of brine shrimp culture to the tubing. Place a cork in the open end of the tubing.

3. When you are ready to count shrimp, divide the tubing into four sections by placing a screw clamp at each mark on the tubing. *While someone holds the corks firmly in place,* first tighten the middle clamp and then the outer clamps.

4. Starting at one end, pour the contents of each section into a test tube labeled with the same number. After you empty a section, loosen the adjacent clamp and fill the next test tube.

5. Stopper one test tube, and invert it gently to distribute the shrimp. Use a pipet to transfer a 1 mL sample of shrimp culture to a Petri dish. Add a few drops of Detain™ to the sample. Count and record the number of live shrimp.

6. Repeat step 5 three more times for the same test tube. Record the average number of shrimp for this test tube.

7. Repeat steps 5 and 6 for each of the remaining test tubes.

PART B: DESIGN AN EXPERIMENT

8. Work with the members of your lab group to explore one of the questions written for step 2 of **Before You Begin.** To explore the question, design an experiment that uses the materials listed for this lab.

You Choose

As you design your experiment, decide the following:

    a. what question you will explore

    b. what hypothesis you will test

    c. how to set up your control

    d. how to expose the brine shrimp to the conditions you chose

    e. how long to expose the brine shrimp to the environmental conditions

    f. how you will set up your data table

Observing How Brine Shrimp Select a Habitat *continued*

9. Write a procedure for your group's experiment. Make a list of all the safety precautions you will take. Have your teacher approve your procedure and safety precautions before you begin the experiment.

10. Set up and conduct your group's experiment. Do *not* use water over 70°C, which can burn you. **CAUTION: If you are working with the hot-water bag, handle it carefully. If you are working with a lamp, do not touch the bulb. Light bulbs get very hot and can burn your skin.**

PART C: CLEANUP AND DISPOSAL

11. Dispose of broken glass in the designated waste container. Put brine shrimp in the designated container. Do not pour chemicals down the drain or put lab materials in the trash unless your teacher tells you to do so.

12. Clean up your work area and all lab equipment. Return lab equipment to its proper place. Wash your hands thoroughly before you leave the lab and after you finish all work.

Analyze and Conclude

1. Summarizing Results Make a bar graph of your data. Use graph paper to plot the environmental variable on the *x*-axis and the number of shrimp on the *y*-axis.

2. Analyzing Results How did the shrimp react to changes in the environment?

_____

3. Analyzing Methods Why was a control necessary?

_____

_____

4. Analyzing Methods Why was it necessary to take many counts in each test tube (step 6 of Part A)?

_____

_____

5. Further Inquiry Write a new question about brine shrimp that could be explored with another investigation.

_____

_____

Skills Practice Lab

Studying Population Growth

SKILLS

- Using a microscope
- Collecting, graphing, and analyzing data
- Calculating

OBJECTIVES

- **Observe** the growth and decline of a population of yeast cells.
- **Determine** the carrying capacity of a yeast culture.

MATERIALS

- safety goggles
- lab apron
- yeast culture
- (2) 1 mL pipets
- 2 test tubes
- 1% methylene blue solution
- ruled microscope slide (2 × 2 mm)
- coverslip
- compound microscope

SAFETY

 CAUTION: Always wear safety goggles and a lab apron to protect your eyes and clothing.

 CAUTION: Do not touch or taste any chemicals. Know the location of the emergency shower and eyewash station and how to use them. If you get a chemical on your skin or clothing, wash it off at the sink while calling to the teacher. Notify the teacher of a spill. Spills should be cleaned up promptly, according to your teacher's directions.

 CAUTION: Glassware is fragile. Notify the teacher of broken glass or cuts. Do not clean up broken glass or spills with broken glass unless the teacher tells you to do so.

Before You Begin

Recall that population size is controlled by **limiting factors**—environmental resources such as food, water, oxygen, light, and living space. **Population growth** occurs when a population's **birth rate** is greater than its **death rate**. A decline in population size occurs when a population's death rate surpasses its birth rate. In this lab, you will study the concepts of population growth, decline, and carrying capacity by growing and observing yeast.

1. Write a definition for each boldface term in the preceding paragraph. Use a separate sheet of paper.

2. You will be using the data table provided on the next page to record your data.

3. Based on the objectives for this lab, write a question about population growth that you would like to explore.

_____

_____

Procedure

PART A: COUNTING YEAST CELLS

1. Put on safety goggles and a lab apron.

2. Transfer 1 mL of a yeast culture to a test tube. Add 2 drops of methylene blue to the tube. **Caution: Methylene blue will stain your skin and clothing.** The methylene blue will remain blue in dead cells but will turn colorless in living cells.

3. Make a wet mount by placing 0.1 mL (one drop) of the yeast and methylene blue mixture on a ruled microscope slide. Cover the slide with a coverslip.

4. Observe the wet mount under the low power of a compound microscope. Notice the squares on the slide. Then switch to the high power. *Note: Adjust the light so that you can clearly see both stained and unstained cells.* Move the slide so that the top left-hand corner of one square is in the center of your field of view. This will be area 1, as shown in the diagram below.

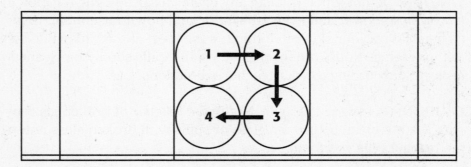

Studying Population Growth *continued*

5. Count the live (unstained) cells and the dead (stained) cells in the four corners of a square using the pattern shown in the diagram. In the data table below, record the numbers of live cells and dead cells in the square.

_____

Data Table			
Time (hours)	Number of cells per square		Population size (cells/0.1 mL)
	Squares 1–6	Average	
0			
24			
48			
72			
96			

6. Repeat step 5 until you have counted 6 squares on the slide. Complete Part B.

_____

7. Find the total number of live cells in the 6 squares. Divide this total by 6 to find the average number of live cells per square. Record this number in the data table. Repeat this procedure for dead cells.

_____

8. Estimate the population of live yeast cells in 1 mL (the amount in the test tube) by multiplying the average number of cells per square by 2,500. Record this number in the data table. Repeat this procedure for dead cells.

_____

9. Repeat steps 1 through 8 each day for 4 more days.

_____

PART B: CLEANUP AND DISPOSAL

10. Dispose of solutions and broken glass in the designated waste containers. Do not pour chemicals down the drain or put lab materials in the trash unless your teacher tells you to do so.

11. Clean up your work area and all lab equipment. Return lab equipment to its proper place. Wash your hands thoroughly before you leave the lab and after you finish all work.

Studying Population Growth *continued*

Analyze and Conclude

1. **Analyzing Methods** Why were several areas and squares counted and then averaged each day?

_____

_____

_____

2. **Summarizing Results** Use graph paper to graph the changes in the numbers of live yeast cells and dead yeast cells over time. Plot the number of cells in 1 mL of yeast culture on the *y*-axis and the time (in hours) on the *x*-axis.

3. **Inferring Conclusions** What limiting factors probably caused the yeast population to decline?

_____

_____

_____

4. **Further Inquiry** Write a new question about population growth that could be explored in another investigation.

_____

_____

Name _____ Class _____ Date _____

Staining and Observing Bacteria

SKILLS

- Using aseptic techniques
- Using a microscope

OBJECTIVES

- **Prepare** and stain wet mounts of bacteria.
- **Identify** different types of bacteria by their shape.

MATERIALS

- wax pencil
- 3 microscope slides
- safety goggles
- lab apron
- disposable gloves
- rubbing alcohol
- paper towels
- 3 culture tubes of bacteria (A, B, and C)
- test-tube rack

- sterile cotton swabs
- Bunsen burner with striker
- microscope slide
- forceps or wooden alligator-type clothespin
- 150 mL beaker
- methylene blue stain in dropper bottle
- compound microscope

SAFETY

 CAUTION: Always wear safety goggles and a lab apron to protect your eyes and clothing.

 CAUTION: Do not touch or taste any chemicals. Know the location of the emergency shower and eyewash station and how to use them. If you get a chemical on your skin or clothing, wash it off at the sink while calling to the teacher. Notify the teacher of a spill. Spills should be cleaned up promptly, according to your teacher's directions.

 CAUTION: Glassware is fragile. Notify the teacher of broken glass or cuts. Do not clean up broken glass or spills with broken glass unless the teacher tells you to do so.

| Staining and Observing Bacteria *continued*

Before You Begin

Like all **prokaryotes,** bacteria are unicellular organisms that sometimes form filaments or loose clusters of cells. They are prepared for viewing by making a **smear,** a slide on which cells have been spread and dried. Treating the cells with a **stain** makes them more visible under magnification. In this lab, you will stain, identify, and compare and contrast different types of bacteria.

1. Write a definition for each boldface term in the paragraph above and for each of the following terms: strepto, staphylo, coccus, bacillus, spirillum. Use a separate sheet of paper.

2. Based on the objectives for this lab, write a question you would like to explore about different kinds of bacteria.

_____

_____

Procedure
PART A: OBSERVING LIVE BACTERIA

1. Put on safety goggles, a lab apron, and disposable gloves.

2. Use a wax pencil to label three microscope slides *A*, *B*, and *C*.

3. Use rubbing alcohol and paper towels to clean the surface of your lab table and gloves. Allow the table to air dry. **CAUTION: Alcohol is flammable. Do not use alcohol near an open flame.**

4. Light a Bunsen burner with a striker. **CAUTION: Keep combustibles away from flames. Do not light a Bunsen burner when others in the room are using alcohol.**

5. Beginning with culture A, make a smear of three different bacteria (A, B, and C) as follows. Remove the cap from a culture tube. *Note: Do not place the cap on the table.* Pass the opening of the tube through the flame of a Bunsen burner. Insert a sterile cotton swab into the tube, and lightly touch the tip of the swab to the bacteria in the culture. Pass the opening of the tube through the flame again, and replace the cap. Transfer a small amount of bacteria to the appropriately labeled microscope slide by rubbing the swab on the slide. Dispose of the swab in a proper container. Repeat for cultures B and C.

6. Allow your smears to air dry. Using microscope slide forceps, pick up each slide one at a time and pass it over the flame several times. Let each slide cool.

80

7. Using microscope slide forceps, place one of your slides across the mouth of a 150 mL beaker half-filled with water. Place 2–3 drops of methylene blue stain on the dried bacteria. *Note: Do not allow the stain to spill into the beaker.* **CAUTION: Methylene blue will stain your skin and clothing.** Let the stain stay on the slide for 2 minutes. Then dip the slide into the water in the beaker several times to rinse it. Blot the slide dry with a paper towel. *Note: Do not rub the slide.*

8. Repeat step 7 for your other two slides.

9. Allow each slide to air dry, and then observe them with a microscope. Make a sketch of a few cells on each slide. Identify the type of bacteria on each slide.

PART B: CLEANUP AND DISPOSAL

10. Dispose of slides, used swabs, solutions, and broken glass in the designated waste containers. Do not pour chemicals down the drain or put lab materials in the trash unless your teacher tells you to do so.

11. Clean up your work area and all lab equipment. *Clean the surface of your lab table with rubbing alcohol.* Return lab equipment to its proper place. Wash your hands thoroughly before you leave the lab and after you finish all work.

Analyze and Conclude

1. Summarizing Results Describe the shape and grouping of the cells of each type of bacteria you observed.

_____

_____

_____

_____

_____

2. Analyzing Methods Why should the test-tube caps from the culture tubes (in Part B) not be placed on the table?

_____

_____

_____

_____

Staining and Observing Bacteria *continued*

3. Evaluating Viewpoints Evaluate the following advice: Always use caution when handling bacteria, even if the bacteria is known to be harmless.

_____

_____

_____

_____

4. Drawing Conclusions How did you classify the bacteria in cultures A, B, and C—as a coccus, a bacillus, or a spirillum?

_____

_____

5. Further Inquiry Write a new question about bacteria that could be explored with another investigation.

_____

_____

Exploration Lab

Observing Protistan Responses to Light

SKILLS

- Using scientific methods
- Using a microscope

OBJECTIVES

- **Identify** several different types of protists.
- **Compare** the structures, methods of locomotion and feeding, and behaviors of several different protists.
- **Relate** a protist's response to light to its method of feeding.

MATERIALS

- Detain™ (protist-slowing agent)
- microscope slides
- plastic pipets with bulbs
- assorted cultures of protists
- toothpicks
- coverslips
- compound microscope
- protist references
- black construction paper
- scissors
- paper punch
- white paper
- sunlit window sill or lamp
- forceps

Before You Begin

Protists belong to the kingdom **Protista**, which is a diverse group of **eukaryotes** that cannot be classified as animals, plants, or fungi. Many protists are unicellular. Among the protists, there are **producers, consumers,** and **decomposers.** In this lab, you will observe live protists and compare their structures, methods of locomotion and feeding, and behaviors.

1. Write a definition for each boldface term in the paragraph above and for each of the following terms: cilia, flagellum, pseudopod. Use a separate sheet of paper.

2. You will be using the data table provided to record your data.

Observing Protistan Responses to Light *continued*

3. Based on the objectives for this lab, write a question you would like to explore about protists.

_____

Procedure
PART A: MAKE OBSERVATIONS

1. Caution: Do not touch your face while handling microorganisms. Place a drop of Detain™ on a microscope slide. Add a drop of liquid from the bottom of a mixed culture of protists. Mix the drops with a toothpick. Add a coverslip. View the slide under low power of a microscope. Switch to high power.

2. Use references to identify the protists. Record data for each type of protist.

Data Table				
Protist	**Color**	**Method of locomotion**	**Method of feeding**	**Other observations**

3. Repeat step 1 *without* using Detain™.

4. Punch a hole in a 40 × 20 mm piece of black construction paper that has a slight curl. The paper should be curled along its long axis with the hole in the middle.

5. Place a wet mount of protists on a piece of white paper. Then put the paper and slide on a sunlit window sill or under a table lamp. Position the sun shade on top of the slide so that the hole is in the center of the coverslip.

6. To examine a slide, first view the area in the center of the hole under low power. *Note: Do not disturb the sun shade. Do not switch to high power.* Then have a partner carefully remove the sun shade with forceps while you observe the slide.

Observing Protistan Responses to Light *continued*

PART B: DESIGN AN EXPERIMENT

7. Work with members of your lab group to explore one of the questions written for step 3 of **Before You Begin.** To explore the question, design an experiment that uses the materials listed for this lab.

You Choose

As you design your experiment, decide the following:

a. what question you will explore

b. what hypothesis you will test

c. how long you will expose protists to light

d. how many times you will repeat your experiment

e. what your control will be

f. what data you will record and how you will make your data table

---

8. Write a procedure for your experiment. Make a list of all the safety precautions you will take. Have your teacher approve your procedure and safety precautions before you begin the experiment.

9. Set up and carry out your experiment.

PART C: CLEANUP AND DISPOSAL

10. Dispose of lab materials and broken glass in the designated waste containers. Put protists in the designated containers. Do not put lab materials in the trash unless your teacher tells you to do so.

11. Clean up your work area and all lab equipment. Return lab equipment to its proper place. Wash your hands thoroughly before you leave the lab and after you finish all work.

Analyze and Conclude

1. Summarizing Results Describe the different types of locomotion you observed in protists, and give examples of each.

_____

_____

_____

_____

_____

Observing Protistan Responses to Light *continued*

2. Analyzing Results Identify which protists were affected by light, and describe how they were affected.

_____

_____

_____

_____

3. Inferring Conclusions What is the relationship between a protist's response to light and its method of feeding?

_____

_____

4. Further Inquiry Write a new question about protists that could be explored with another investigation.

_____

_____

_____

Skills Practice Lab

Observing Yeast and Fermentation

SKILLS

- Observing
- Measuring
- Collecting data
- Analyzing data

OBJECTIVE

- **Observe** the release of energy by yeast during fermentation.

MATERIALS

- 500 mL vacuum bottle
- 10 cm glass tubing
- 2-hole rubber stopper
- 250 mL beaker
- 75 g sucrose
- one package dry yeast
- thermometer
- 50 cm rubber tubing
- 150 mL limewater

SAFETY

  CAUTION: Always wear safety goggles and a lab apron to protect your eyes and clothing.

 CAUTION: Glassware is fragile. Notify the teacher of broken glass or cuts. Do not clean up broken glass or spills with broken glass unless the teacher tells you to do so.

Before You Begin

Sucrose is a disaccharide—a carbohydrate made from two monosaccharides. It is one chemical made by plants to store the sun's energy. Yeast release the energy stored in sucrose in a process called **fermentation.** In this investigation you will have a chance to observe and measure the products of fermentation.

1. Write a definition for the boldface term in the paragraph above.

_____

_____

Observing Yeast and Fermentation *continued*

2. You will be using the data table below to record your data.

Data Table								
Fermentation by Yeast								
Time	**Date**	**Temp.**	**Time**	**Date**	**Temp.**	**Time**	**Date**	**Temp.**
1.			8.			15.		
2.			9.			16.		
3.			10.			17.		
4.			11.			18.		
5.			12.			19.		
6.			13.			20.		
7.			14.			21.		

Procedure

1. Set up your vacuum bottle according to the diagram.

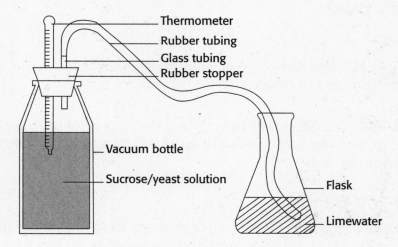

2. Mix 75 g of sucrose in 400 mL of water.

3. When the sucrose has dissolved, add one-half package of fresh yeast and stir.

4. Pour the sucrose-yeast solution into a vacuum bottle until it is approximately three-quarters full.

5. Adjust the thermometer so that it extends down into the sugar-yeast solution.

Observing Yeast and Fermentation *continued*

6. Record the temperature of the solution on the observation chart as soon as possible. Continue to record the temperature as often as possible during the next two days.

Analyze and Conclude

1. Summarizing Results Prepare a graph of your data, illustrating the temperature over time. Complete the graph by drawing a curve through the plotted points.

_____

_____

2. Analyzing Data What does the curved line plotted on the graph indicate?

_____

_____

3. Drawing Conclusions What can you conclude about the energy contained in sucrose?

_____

_____

_____

_____

_____

4. Predicting Patterns What do you think would happen if there were only one hole in the stopper for the thermometer?

_____

_____

5. Further Inquiry If you know that fermentation liberates energy and gives off carbon dioxide and alcohol as waste products, how would you prove that fermentation is really taking place in the sugar-yeast solution?

_____

_____

_____

_____

Skills Practice Lab

Surveying Plant Diversity

SKILLS

- Observing
- Comparing

OBJECTIVES

- **Identify** similarities and differences among four phyla of living plants.
- **Relate** structural adaptations of plants to their success on land.

MATERIALS

- live or preserved specimens of mosses, ferns, conifers, and flowering plants
- stereomicroscope or hand lens
- compound microscope
- prepared slides of fern gametophytes

Before You Begin

Most plants are complex photosynthetic organisms that live on land. The ancestors of plants lived in water. As plants evolved on land, however, they developed adaptations that made it possible for them to be successful in dry conditions. All plant life cycles are characterized by **alternation of generations,** in which a haploid **gametophyte** stage alternates with a diploid **sporophyte** stage. Distinct differences in the relative sizes and structures of gametophytes and sporophytes are seen among the 12 phyla of living plants. In this lab, you will examine representatives of the four most familiar plant phyla.

1. Write a definition for each boldface term in the paragraph above and for the following terms: sporangium, spore, frond, cone, flower, fruit. Use a separate sheet of paper.

2. You will be using the data table provided to record your data.

3. Based on the objectives for this lab, write a question you would like to explore about the characteristics of plants.

_____

Procedure

PART A: CONDUCTING A SURVEY

1. Visit the station for each of the plants, and examine the specimens there. Answer the questions, and record observations in your data table.

Surveying Plant Diversity *continued*

Data Table

Phylum name	Dominant generation	Major characteristics	Examples

2. Mosses Examine a clump of moss with a stereomicroscope or hand lens. Make a sketch of what you see. Use a separate sheet of paper.

3. Mosses Examine a moss gametophyte with a sporophyte attached to it. Draw what you see, and label the parts you recognize. Label each part as haploid or diploid.

a. Which stage of a moss has rootlike structures?

_____

b. Where are the spores of a moss produced?

_____

4. Ferns Examine the sporophyte of a fern, and look for evidence of reproductive structures on the underside of the fronds. Draw what you see. Label a leaf (frond), stem, root, and reproductive structure.

a. How does water travel through a fern? List observations supporting your answer.

_____

_____

b. What kind of reproductive cells are produced by fern fronds?

_____

5. Ferns Examine a slide of a fern gametophyte with a compound microscope. Draw what you see, and label any structures you recognize.

_____

Surveying Plant Diversity *continued*

6. **Conifers** Draw a part of a branch of one of the conifers at this station. Label a leaf, stem, and cone (if present).

   a. Is a branch of a pine tree part of a gametophyte or part of a sporophyte?

   _____

   b. In what part of a conifer would you look to find its reproductive structures?

   _____

7. **Conifers** Examine a prepared slide of pine pollen. Draw a few of the grains.

   a. What reproductive structure is found within a pollen grain?

   _____

   b. How does the structure of pine pollen aid in its dispersal by wind?

   _____

8. **Angiosperms** Draw one of the representative angiosperms at this station. Label a leaf, stem, root, and flower (if present). Indicate the sporophyte and location of gametophytes.

   a. Where do angiosperms produce sperm and eggs?

   _____

   b. How do the seeds of angiosperms differ from those of gymnosperms?

   _____

   _____

9. **Angiosperms** Examine several fruits. Draw and label the parts of one fruit.

PART B: CLEANUP AND DISPOSAL

10. Dispose of broken glass in the designated waste containers. Do not put lab materials in the trash unless your teacher tells you to do so.

11. Wash your hands thoroughly before you leave the lab and after you finish all work.

Analyze and Conclude

1. **Analyzing Information** How are bryophytes different from the other major groups of plants?

   _____

   _____

   _____

Surveying Plant Diversity *continued*

2. Recognizing Patterns How do the gametophytes of gymnosperms and angiosperms differ from the gametophytes of bryophytes and ferns?

_____

_____

_____

3. Drawing Conclusions What structures are present in both gymnosperms and angiosperms but absent in both bryophytes and ferns?

_____

4. Evaluating Hypotheses Dispersal is the main function of fruits in angiosperms. Defend or refute this hypothesis. List observations you made during this lab to support your position.

_____

_____

_____

_____

5. Inferring Conclusions Based on their characteristics, which phylum of plants appears to be the most successful? Justify your conclusion.

_____

_____

_____

_____

_____

6. Further Inquiry Write a new question about plant diversity that could be explored with another investigation.

_____

_____

Exploration Lab

Observing the Effects of Nutrients on Vegetative Reproduction

SKILLS

- Using scientific processes
- Observing
- Graphing and analyzing data

OBJECTIVES

- **Identify** the structures of duckweed.
- **Compare** vegetative reproduction of duckweed in different nutrient solutions.

MATERIALS

- safety goggles
- lab apron
- duckweed culture
- 5 Petri dishes
- stereomicroscope or hand lens
- glass-marking pen
- beakers
- pond water
- Knop's solution
- 0.1% fertilizer solution
- distilled water

SAFETY

 CAUTION: Always wear safety goggles and a lab apron to protect your eyes and clothing.

CAUTION: Do not touch or taste any chemicals. Know the location of the emergency shower and eyewash station and how to use them. If you get a chemical on your skin or clothing, wash it off at the sink while calling to the teacher. Notify the teacher of a spill. Spills should be cleaned up promptly, according to your teacher's directions.

CAUTION: Glassware is fragile. Notify the teacher of broken glass or cuts. Do not clean up broken glass or spills with broken glass unless the teacher tells you to do so.

Observing the Effects of Nutrients on Vegetative Reproduction *continued*

Before You Begin

Duckweed is a common aquatic plant. Like many flowering plants, duckweed reproduces readily by **vegetative reproduction,** which is a type of **asexual reproduction.** As individual plants grow, they divide into smaller individuals. Several individuals may remain joined together, forming a mat. All plants require certain **mineral nutrients,** such as nitrogen, phosphorus, and potassium, for the growth of vegetative parts. In this lab, you will investigate the effect of nutrients on the vegetative reproduction of duckweed.

1. Write a definition for each boldface term in the paragraph above. Use a separate sheet of paper.

2. Based on the objectives for this lab, write a question you would like to explore about vegetative reproduction in duckweed.

_____

_____

Procedure
PART A: MAKE OBSERVATIONS

1. Place a duckweed plant in a Petri dish. Then place a few drops of water on the plant.

2. Observe the duckweed plant with a stereomicroscope or a hand lens. Sketch what you see. Label the structures that you recognize.

PART B: DESIGN AN EXPERIMENT

3. Work with members of your lab group to explore one of the questions written for step 2 of **Before You Begin.** To explore the question, design an experiment that uses the materials listed for this lab.

> **You Choose**
>
> As you design your experiment, decide the following:
>
> **a.** what question you will explore
>
> **b.** what your hypothesis will be
>
> **c.** what solutions to test
>
> **d.** how much of each solution to use
>
> **e.** how many individuals to use for each test
>
> **f.** what your control will be
>
> **g.** how you will judge which solution is the best
>
> **h.** what data to record in your data table

Observing the Effects of Nutrients on Vegetative Reproduction *continued*

4. Write a procedure for your experiment. Make a list of all the safety precautions you will take. Have your teacher approve your procedure and safety precautions before you begin the experiment.

PART C: CONDUCT YOUR EXPERIMENT

5. Put on safety goggles and a lab apron.

6. Set up your experiment. **CAUTION: Nutrient solutions are mild eye irritants. Avoid contact with your skin and eyes.** Complete step 8.

7. Conduct your experiment and collect data for two weeks.

PART D: CLEANUP AND DISPOSAL

8. Dispose of solutions, broken glass, and duckweed in the designated waste containers. Do not pour chemicals down the drain or put lab materials in the trash unless your teacher tells you to do so.

9. Clean up your work area and all lab equipment. Return lab equipment to its proper place. Wash your hands thoroughly before you leave the lab and after you finish all work.

Analyze and Conclude

1. Summarizing Results Compare the appearance of plants growing in each nutrient solution with that of the plants in distilled water. Explain your observations.

_____

_____

_____

_____

2. Analyzing Data In which Petri dish did the greatest amount of growth (increase in numbers) take place?

_____

3. Analyzing Results In which Petri dish did the least amount of growth take place?

_____

4. Evaluating Hypotheses Did the results you observed agree with your hypothesis? If not, how are they different?

_____

_____

Observing the Effects of Nutrients on Vegetative Reproduction *continued*

5. Recognizing Patterns As the number of new duckweed plants in a particular group increased, what happened to the group of plants?

_____

_____

6. Graphing Data Make a graph of your data. Label the *y*-axis "Number of plants," and the *x*-axis "Days." Use a different color to represent each solution you tested.

_____

_____

7. Drawing Conclusions What factors regulate the rate of vegetative reproduction in duckweed?

_____

_____

_____

8. Evaluating Methods Why are the new duckweed plants produced by vegetative reproduction genetically the same as the parent plant?

_____

_____

_____

9. Further Inquiry Write a new question about vegetative reproduction in duckweed that could be explored with another investigation.

_____

_____

Name _____ Class _____ Date _____

Separating Plant Pigments

SKILLS

• Performing paper chromatography

• Calculating

OBJECTIVES

• **Separate** the pigments that give a leaf its color.

• **Calculate** the R_f value for each pigment.

• **Describe** how paper chromatography can be used to study plant pigments.

MATERIALS

• safety goggles

• lab apron

• strip of chromatography paper

• scissors

• metric ruler

• pencil

• capillary tube

• drop of simulated plant pigments extract

• 10 mL graduated cylinder

• 5 mL of chromatography solvent

• chromatography chamber

SAFETY

  CAUTION: Always wear safety goggles and a lab apron to protect your eyes and clothing.

 CAUTION: Do not touch or taste any chemicals. Know the location of the emergency shower and eyewash station and how to use them. If you get a chemical on your skin or clothing, wash it off at the sink while calling to the teacher. Notify the teacher of a spill. Spills should be cleaned up promptly, according to your teacher's directions.

CAUTION: Glassware is fragile. Notify the teacher of broken glass or cuts. Do not clean up broken glass or spills with broken glass unless the teacher tells you to do so.

Before You Begin

Pigments produce colors by reflecting some colors of light and absorbing or transmitting others. Pigments can be removed from plant tissues using **solvents,** chemicals that dissolve other chemicals. The pigments can then be separated from the solvent and from each other by using **paper chromatography.** The word *chromatography* comes from the Greek words *chromat,* which means 'color,' and *graphon,* which means 'to write.' The R_f is the ratio of the distance that a pigment moves relative to the distance that a solvent moves. Since the R_f for a compound is constant, scientists can use it to identify compounds. In this lab, you will learn how to use paper chromatography to separate a mixture of pigments.

1. Write a definition for each boldface term in the previous paragraph and for each of the following terms: **chlorophyll a, chlorophyll b, carotene, xanthophyll.** Use a separate sheet of paper.

2. You will be using the data table provided to record your data.

3. Based on the objectives for this lab, write a question you would like to explore about plant pigments or paper chromatography.

_____

_____

Procedure
PART A: MAKING A CHROMATOGRAM

1. Put on safety goggles and a lab apron. Use scissors to cut the bottom end of a strip of chromatography paper to a tapered end. **CAUTION: Sharp or pointed objects may cause injury. Handle scissors carefully.**

2. Draw a faint pencil line 1 cm above the pointed end of the paper strip. Use a capillary tube to apply a tiny drop of the simulated plant pigments extract on the center of the line.

3. Pour 5 mL of chromatography solvent into a chromatography chamber. Pull the chromatography paper through the opening of the cap, and adjust the length of the strip so that a small portion of the tip end is immersed in the solvent. DO NOT immerse the pigment in the solvent.

4. Place the cap over the chromatography chamber. Carefully bend the end of the strip of chromatography paper over the cap. Be sure that the strip does not touch the walls of the chamber.

5. Remove the strip from the chromatography chamber when the solvent nears the top of the chamber (within 5–7 minutes).

Separating Plant Pigments *continued*

6. With a pencil, mark the position of the uppermost end of the solvent and the farthest distance each pigment moved. Measure the distance that the solvent and each pigment moved. Record your observations and measurements in your data table. Tape or glue your chromatogram to your lab report. Label the pigment colors.

Data Table				
Band no.	**Color**	**Pigment**	**Migration (in mm)**	R_f **value**
1 (top)				
2				
3				
4				
Solvent				

7. Use the formula below to calculate and record the R_f for each pigment.

$$R_f = \frac{Distance\ substance\ (pigment)\ traveled}{distance\ solvent\ traveled}$$

PART B: CLEANUP AND DISPOSAL

8. Dispose of chromatography paper, solutions, and broken glass in the designated waste containers. Do not pour chemicals down the drain or put lab materials in the trash unless your teacher tells you to do so.

9. Clean up your work area and all lab equipment. Return lab equipment to its proper place. Wash your hands thoroughly before you leave the lab and after you finish all work.

Analyze and Conclude

1. Summarizing Results Describe what happened to the simulated plant pigments during the lab.

_____

_____

_____

_____

_____

Separating Plant Pigments *continued*

2. Analyzing Data How do your R_f values compare with those of your classmates?

_____

_____

_____

3. Inferring Conclusions What is a chromatogram?

_____

_____

_____

4. Further Inquiry Write a new question about plant pigments that could be explored with another investigation.

_____

_____

_____

Skills Practice Lab

Comparing Bean and Corn Seedlings

SKILLS

- Comparing
- Drawing
- Relating

OBJECTIVES

- **Observe** the structures of bean seeds and corn kernels.
- **Compare** and **contrast** the development of bean embryos and corn embryos as they grow into seedlings.

MATERIALS

- 6 bean seeds soaked overnight
- stereomicroscope
- 6 corn kernels soaked overnight
- scalpel
- paper towels
- 2 rubber bands
- 150 mL beakers (2)
- glass-marking pen
- metric ruler

Before You Begin

A **seed** contains an inactive plant **embryo.** A plant embryo consists of one or more **cotyledons,** an embryonic shoot, and an embryonic **root.** Seeds also contain a supply of nutrients. In **monocots,** the nutrients are contained in the **endosperm.** In **dicots,** the nutrients are transferred to the cotyledons as seeds mature. A seed **germinates** when the embryo begins to grow and breaks through the protective **seed coat.** The embryo then develops into a young plant, or **seedling.** In this lab, you will examine bean seeds and corn kernels and then germinate them to observe the development of their seedlings.

1. Write a definition for each boldface term in the paragraph above. Use a separate sheet of paper.

2. Based on the objectives for this lab, write a question you would like to explore about seedling development.

_____

_____

Comparing Bean and Corn Seedlings *continued*

Procedure

PART A: OBSERVING SEED STRUCTURE

1. Remove the seed coat of a bean seed, and separate the two fleshy halves of the seed.

2. Locate the embryo on one of the halves of the seed. Examine the bean embryo with a stereomicroscope. Draw the embryo, and label the parts you can identify.

3. 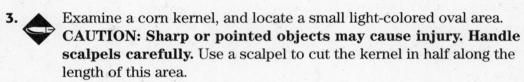 Examine a corn kernel, and locate a small light-colored oval area. **CAUTION: Sharp or pointed objects may cause injury. Handle scalpels carefully.** Use a scalpel to cut the kernel in half along the length of this area.

4. Locate the corn embryo, and examine it with a stereomicroscope. Draw the embryo, and label the parts you can identify.

PART B: OBSERVING SEEDLING DEVELOPMENT

5. Fold a paper towel in half. Set five corn kernels on the paper towel. Roll up the paper towel, and put a rubber band around the roll. Stand the roll in a beaker with 1 cm of water in the bottom. Add water to the beaker as needed to keep the paper towels wet, but do not allow the corn kernels to be covered by water.

6. Repeat step 5 with five bean seeds.

7. After three days, unroll the paper towels and examine the corn and bean seedlings. Use a glass-marking pen to mark the roots and shoots of the developing seedlings. Starting at the seed, make a mark every 0.5 cm along the root of each seedling. And again starting at the seed, make a mark every 0.5 cm along the stem of each seedling.

Comparing Bean and Corn Seedlings *continued*

8. Draw a corn seedling and a bean seedling in your lab report. Label the parts of each seedling. Also show the marks you made on each seedling, and indicate the distance between the marks.

9. Using a fresh paper towel, roll up the seeds, place the rolls in the beakers, and add fresh water to the beakers.

10. After two more days reexamine the seedlings. Measure the distance between the marks. Repeat step 8.

PART C: CLEANUP AND DISPOSAL

11. Dispose of seeds, broken glass, and paper towels in the designated waste containers. Do not put lab materials in the trash unless your teacher tells you to do so.

12. Clean up your work area and all lab equipment. Return lab equipment to its proper place. Wash your hands thoroughly before you leave the lab and after you finish all work.

Analyze and Conclude

1. **Relating Concepts** Corn and beans are often cited as representative examples of monocots and dicots, respectively. Relate the seed structure of each to the terms *monocotyledon* and *dicotyledon*.

_____

_____

_____

_____

2. **Summarizing Results** What parts of a plant embryo were observed in all seedlings on the third day?

_____

_____

Comparing Bean and Corn Seedlings *continued*

3. Drawing Conclusions In which part or parts of bean seedlings and corn seedlings do the seedlings grow in length? Explain.

_____

_____

_____

_____

_____

_____

4. Forming Hypotheses How are the tender young shoots of bean seedlings and corn seedlings protected as the seedlings grow through the soil?

_____

_____

_____

_____

_____

_____

5. Evaluating Viewpoints Defend the following statement: There are both similarities and differences in seed structure and seedling development in beans and corn.

_____

_____

_____

_____

_____

6. Further Inquiry Write a new question about seedling development that could be explored with another investigation.

_____

_____

_____

Surveying Invertebrate Diversity

SKILLS

- Observing
- Comparing

OBJECTIVES

- **Observe** the similarities and differences among groups of invertebrates.
- **Relate** the structural adaptations of invertebrates to their evolution.

MATERIALS

- safety goggles
- lab apron
- preserved or living specimens of invertebrates
- prepared slides of sponges, hydras, planarians, and nematodes
- compound microscopes
- hand lenses or stereomicroscopes
- probes

SAFETY

 CAUTION: Always wear safety goggles and a lab apron to protect your eyes and clothing.

CAUTION: Do not touch or taste any chemicals. Know the location of the emergency shower and eyewash station and how to use them. If you get a chemical on your skin or clothing, wash it off at the sink while calling to the teacher. Notify the teacher of a spill. Spills should be cleaned up promptly, according to your teacher's directions.

CAUTION: Glassware is fragile. Notify the teacher of broken glass or cuts. Do not clean up broken glass or spills with broken glass unless the teacher tells you to do so.

Before You Begin

Invertebrates include all animals except those with backbones. Every phylum of the kingdom Animalia except the phylum Chordata consists only of invertebrates. In this lab, you will examine representatives of eight phyla of animals.
You will see many similarities and differences in **body plan**—shape, symmetry, and internal organization.

1. Write a definition for each boldface term in the paragraph above and for the following terms: radial symmetry, bilateral symmetry, dorsal, ventral, anterior, posterior, cephalization, segmentation. Use a separate sheet of paper.

2. Describe the three basic body plans found in animals. Use a separate sheet of paper.

3. You will be using the data table provided to record your data.

4. Based on the objectives for this lab, write a question you would like to explore about the characteristics of invertebrates.

_____

_____

Procedure

PART A: CONDUCTING A SURVEY

1. Put on safety goggles and a lab apron.

2. Visit each invertebrate station, and examine the specimens there. Answer the questions, and record observations in your data table.

Data Table				
Phylum	**Symmetry**	**Body plan**	**Other**	**Examples**

3. Sponges Examine each specimen.

a. Describe the shape of a sponge.

_____

b. What do you think is the role of the many holes, or pores, in a sponge?

_____

_____

c. Examine a prepared slide of a sponge with a compound microscope. What do you notice about the organization of the cells in sponges?

_____

_____

Surveying Invertebrate Diversity *continued*

4. Cnidarians Examine each specimen.

a. Divide the cnidarian specimens into two groups. What feature did you use to make your division?

_____

b. How many body openings does a cnidarian have?

_____

c. Examine a prepared slide of a hydra. What do you notice about the organization of the cells in cnidarians?

_____

5. Flatworms and Roundworms Examine each specimen.

a. How does a flatworm differ from a roundworm in external appearance?

_____

_____

_____

b. Do any of the worms appear to be segmented? Explain.

_____

_____

_____

_____

c. Examine prepared slides of planarians and nematodes. How many body openings does each have?

_____

6. Mollusks Examine each specimen.

a. In what ways do the mollusks differ in external appearance?

_____

b. Which group of mollusks has the most noticeable "feet"?

_____

Surveying Invertebrate Diversity *continued*

7. Annelids Examine each specimen.

   a. How are an earthworm and a leech similar? How are they different?

_____

_____

_____

_____

   b. Describe any differences you see in the segments of the annelid worm.

_____

_____

8. Arthropods Examine each specimen.

   a. What characteristic do you observe in all arthropod appendages?

_____

   b. How does the number of walking legs differ among these arthropods?

_____

_____

_____

9. Echinoderms Examine each specimen.

   a. The word *echinoderm* means "spiny skin." Why is this name appropriate?

_____

_____

   b. What does an echinoderm's ventral surface look like?

_____

_____

PART B: CLEANUP AND DISPOSAL

10. Dispose of broken glass in the designated waste containers. Do not put lab materials in the trash unless your teacher tells you to do so.

11. Wash your hands thoroughly before you leave the lab and after you finish all work.

Surveying Invertebrate Diversity *continued*

Analyze and Conclude

1. Summarizing Data Which animal phyla show cephalization, and which do not?

_____

_____

2. Recognizing Patterns What type of symmetry is found with cephalization?

_____

3. Recognizing Patterns What characteristics do annelids and arthropods share?

_____

_____

4. Analyzing Methods Were you able to identify the type of body plan found in all of the specimens? Explain.

_____

_____

5. Further Inquiry Write a new question about invertebrates that could be explored with another investigation.

_____

_____

Exploration Lab

Observing Hydra Behavior

SKILLS

• Using scientific processes

• Observing

OBJECTIVES

• **Observe** a hydra finding and capturing prey.

• **Determine** how a hydra responds to stimuli.

MATERIALS

• silicone culture gum

• microscope slide

• 2 medicine droppers

• *Hydra* culture

• *Daphnia* culture

• concentrated beef broth

• filter paper cut into pennant shapes

• forceps

• stereomicroscope

Before You Begin

Cnidarians are carnivorous animals. A common cnidarian is **Hydra**, a freshwater organism that feeds on smaller freshwater animals, such as water fleas (*Daphnia*). Hydras find food by responding to stimuli, such as chemicals and touch. The way an animal responds to stimuli is called **behavior.** The tentacles of a cnidarian are armed with **nematocysts,** which are used in defense and in capturing prey. When a hydra receives stimuli from potential prey, its nematocysts spring out and harpoon or entangle the prey. In this lab, you will observe the feeding behavior of hydras to determine how they find and capture prey.

1. Write a definition for each boldface term in the paragraph above. Use a separate sheet of paper

2. Based on the objectives for this lab, write a question you would like to explore about the feeding behavior of hydras.

_____

_____

Procedure

PART A: MAKE OBSERVATIONS

1. To make an experimental pond for observing hydras, squeeze out a long piece of silicone culture gum. Arrange it to form a circular well on a microscope slide. **CAUTION: Glassware is fragile. Notify the teacher promptly of any broken glass or cuts.**

2. With a medicine dropper, gently transfer a hydra from its culture dish to the well on the slide, making sure the water covers the animal. **CAUTION: Handle hydras gently to avoid injuring them.** Allow the hydra to settle, then examine it under the high power of a stereomicroscope. Draw a hydra and label the body stalk, mouth, and tentacles.

PART B: DESIGN AN EXPERIMENT

3. Work with the members of your lab group to explore one of the questions written for step 2 of **Before You Begin.** To explore the question, design an experiment that uses the materials listed for this lab.

4. Write a procedure for your experiment. Make a list of all the safety precautions you will take. Have your teacher approve your procedure and safety precautions before you begin the experiment.

You Choose

As you design your experiment, decide the following:

a. what question you will explore

b. what hypothesis you will test

c. how to observe a hydra's feeding behavior

d. how to test a hydra's response to a stimulus, such as a chemical or a touch

e. what your test groups and controls will be

f. what to record in your data table

PART C: CONDUCT YOUR EXPERIMENT

5. Set up and carry out your experiment. **CAUTION: Handle hydras gently to avoid injuring them.**

6. Allow hydras to settle before exposing them to a test condition. If your hydra does not respond after a few minutes, obtain another hydra from the culture dish. Repeat your procedure.

Observing Hydra Behavior *continued*

PART D: CLEANUP AND DISPOSAL

7. Dispose of lab materials and broken glass in the designated waste containers. Put hydras and daphnias in the designated containers. Do not put lab materials in the trash unless your teacher tells you to do so.

8. Clean up your work area and all lab equipment. Return lab equipment to its proper place. Wash your hands thoroughly before you leave the lab and after you finish all work.

Analyze and Conclude

1. Analyzing Results Describe a hydra's response to chemicals (beef broth).

_____

_____

_____

2. Analyzing Results Describe a hydra's response to touch.

_____

_____

3. Drawing Conclusions How does a hydra detect its prey?

_____

_____

4. Justifying Conclusions Give evidence to support your conclusion about how hydras detect prey.

_____

_____

_____

5. Inferring Conclusions Based on your observations, how do you think a hydra behaves when it detects a threat in its natural habitat?

_____

_____

_____

6. Inferring Conclusions What happens to food that has not been digested by a hydra?

_____

_____

Observing Hydra Behavior *continued*

7. Inferring Conclusions How is a hydra adapted to a sedentary lifestyle?

_____

_____

_____

_____

8. Further Inquiry Write a new question about the behavior of hydras that could be explored with another investigation.

_____

_____

Observing Characteristics of Clams

SKILLS

- Observing
- Testing for the presence of a chemical

OBJECTIVES

- **Observe** the behavior of a live clam.
- **Examine** the structure and composition of a clam shell.

MATERIALS

- safety goggles
- lab apron
- live clam
- small beaker or dish
- eyedropper
- food coloring
- glass stirring rod
- clam shell
- Petri dish
- scalpel
- stereomicroscope
- 0.1 M HCl

SAFETY

  CAUTION: Always wear safety goggles and a lab apron to protect your eyes and clothing.

 CAUTION: Do not touch or taste any chemicals. Know the location of the emergency shower and eyewash station and how to use them. If you get a chemical on your skin or clothing, wash it off at the sink while calling to the teacher. Notify the teacher of a spill. Spills should be cleaned up promptly, according to your teacher's directions.

 CAUTION: Glassware is fragile. Notify the teacher of broken glass or cuts. Do not clean up broken glass or spills with broken glass unless the teacher tells you to do so.

| Observing Characteristics of Clams *continued*

Before You Begin

Clams are **mollusks,** and they have a two-part shell. The body of a clam consists of a visceral mass and a muscular **foot.** There is no definite head. Two tubes, an **incurrent siphon** and an **excurrent siphon,** extend from the body on the side opposite the foot. Like all mollusks, clams have a shell composed of **calcium carbonate.** A membrane called the **mantle** lines the shell and forms successive rings of shell as a clam grows. The **umbo** is the oldest part of a clam shell. In this lab, you will examine live clams and clam shells.

1. Write a definition for each boldface term in the paragraph. Use a separate sheet of paper.

2. Based on your objectives, write a question you would like to explore about clams.

_____

_____

Procedure

PART A: OBSERVE A LIVE CLAM

1. Put on safety goggles and a lab apron.

2. Place a live clam in a small beaker or shallow dish of water. Using an eyedropper, apply two drops of food coloring near the clam.

3. Observe and record what happens to the food coloring.

_____

_____

4. Using a stirring rod, touch the clam's mantle. **CAUTION: Touch the clam gently to avoid injuring it.**

5. Observe and record the clam's response to touch.

_____

_____

❚ Observing Characteristics of Clams *continued*

PART B: OBSERVE A CLAM SHELL

6. Examine the concentric growth rings on the shell. Locate the knob-shaped umbo on the shell. Count and record the number of growth rings on the clam shell.

_____

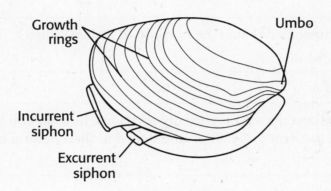

7. Place the clam shell in a Petri dish. Use a scalpel to chip away part of the shell to expose its three layers. **CAUTION: Sharp or pointed objects may cause injury. Handle scalpels carefully.** View the shell's layers with a stereomicroscope. The outermost layer protects the clam from acids in the water. The innermost layer is mother-of-pearl, the material that forms pearls.

8. The middle layer of the shell contains crystals of calcium carbonate. To test for the presence of this compound, place one drop of 0.1 M HCl on the middle layer of the shell. **CAUTION: Hydrochloric acid is corrosive. Avoid contact with skin, eyes, and clothing. Avoid breathing vapors.** If calcium carbonate is present, bubbles of carbon dioxide will form in the drop. Record your observations.

_____

_____

PART C: CLEANUP AND DISPOSAL

9. Dispose of solutions, broken glass, and pieces of clam shell in the waste containers designated by your teacher. Do not pour chemicals down the drain or put lab materials in the trash unless your teacher tells you to do so.

10. Clean up your work area and all lab equipment. Return live clams to the stock container. Return lab equipment to its proper place. Wash your hands thoroughly before you leave the lab and after you finish all work.

Analyze and Conclude

1. **Analyzing Results** Find the incurrent and excurrent siphons of the clam in the illustration on the previous page. Using this information, explain your observations in step 3.

_____

_____

2. **Drawing Conclusions** What is the purpose of a clam's shell?

_____

_____

3. **Making Predictions** Based on your observations, how do you think clams respond when they are touched or threatened in their natural habitat?

_____

_____

4. **Forming a Hypothesis** What does a clam take in from water that passes through its body?

_____

5. **Inferring Relationships** Water that enters a clam's incurrent siphon passes over the clam's gills. How does this help the clam respire?

_____

_____

6. **Further Inquiry** Write a new question about clams that could be explored with another investigation.

_____

_____

Observing Pill Bug Behavior

SKILLS

- Using scientific methods
- Observing

OBJECTIVES

- **Identify** arthropod characteristics in a pill bug.
- **Observe** the behavior of pill bugs on surfaces with different textures.
- **Infer** the adaptive advantages of pill bug behaviors.

MATERIALS

- 4 adult pill bugs
- 2 Petri dishes
- stereomicroscope or hand lens
- blunt probe
- fabrics with different textures
- scissors
- transparent tape
- clock or watch with second hand

SAFETY

  CAUTION: Always wear safety goggles and a lab apron to protect your eyes and clothing.

 CAUTION: Do not touch or taste any chemicals. Know the location of the emergency shower and eyewash station and how to use them. If you get a chemical on your skin or clothing, wash it off at the sink while calling to the teacher. Notify the teacher of a spill. Spills should be cleaned up promptly, according to your teacher's directions.

 CAUTION: Glassware is fragile. Notify the teacher of broken glass or cuts. Do not clean up broken glass or spills with broken glass unless the teacher tells you to do so.

Before You Begin

Pill bugs live in moist terrestrial environments, such as under rocks and logs. Like other **crustaceans,** pill bugs respire with gills and have hard outer shells and jointed appendages. They respond to a **stimulus,** such as light, moisture, or touch, by moving toward or away from the stimulus or by curling into a ball. In this lab, you will look for arthropod characteristics in pill bugs and observe the behavior of pill bugs on surfaces with different textures.

Observing Pill Bug Behavior *continued*

1. Write a definition for each boldface term in the preceding paragraph. Use a separate sheet of paper.

2. Based on the objectives for this lab, write a question you would like to explore about pill bug characteristics and behavior.

_____

_____

Procedure

PART A: MAKE OBSERVATIONS

1. Place a pill bug in a Petri dish, and observe it with a stereomicroscope or hand lens. Observe it from a dorsal viewpoint as well as from the side. List the characteristics that tell you the pill bug is an arthropod.

_____

_____

_____

2. Touch the pill bug with a blunt probe. **CAUTION: Touch pill bugs gently to avoid injuring them.** Record your observations.

_____

_____

_____

PART B: DESIGN AN EXPERIMENT

3. Work with the members of your lab group to explore one of the questions written for step 2 of **Before You Begin.** To explore the question, design an experiment that uses the materials listed for this lab.

> **You Choose**
>
> As you design your experiment, decide the following:
>
> **a.** what question you will explore
>
> **b.** what hypothesis you will test
>
> **c.** which four different fabrics to use
>
> **d.** how many times to test each fabric
>
> **e.** the length of each test
>
> **f.** what your control will be
>
> **g.** what data to record in your data table

4. Write a procedure for your experiment. Make a list of all the safety precautions you will take. Have your teacher approve your procedure and safety precautions before you begin the experiment.

5. Set up and carry out your experiment. **CAUTION: Sharp or pointed objects can cause injury. Handle scissors carefully.**

PART C: CLEANUP AND DISPOSAL

6. Dispose of fabric scraps and broken glass in the designated waste containers. Put pill bugs in the designated container. Do not put lab materials in the trash unless your teacher tells you to do so.

7. Clean up your work area and all lab equipment. Return lab equipment to its proper place. Wash your hands thoroughly before you leave the lab and after you finish all work.

Analyze and Conclude

1. Analyzing Methods Why did you test several pill bugs in this investigation instead of just one pill bug?

_____

_____

2. Analyzing Results Did all of your pill bugs show a similar pattern of movement? Explain.

_____

_____

3. Graphing Results Make a graph of your data on a sheet of graph paper. Plot the average time spent on the material on the *y*-axis and the type of material on the *x*-axis.

4. Analyzing Results Rank the fabrics according to the total amount of time spent on them by the pill bugs.

_____

_____

_____

5. Drawing Conclusions Which fabric texture do pill bugs seem to prefer?

_____

_____

_____

Observing Pill Bug Behavior *continued*

6. Inferring Conclusions How is a pill bug's response to disturbances an advantage?

_____

_____

_____

7. Inferring Conclusions How is being able to detect surface texture helpful to pill bugs in their natural habitat?

_____

_____

_____

8. Further Inquiry Write a new question about pill bugs that could be explored with another investigation.

_____

_____

_____

Skills Practice Lab

Analyzing Sea Star Anatomy

SKILLS

- Observing
- Collecting data
- Inferring

OBJECTIVES

- **Observe** anatomical structures of an echinoderm.
- **Infer** function of body parts from structure.

MATERIALS

- disposable gloves
- preserved sea star
- dissection tray
- dissection scissors
- hand lens
- dissecting microscope
- forceps
- blunt probe
- sharp probe
- dissection pins

SAFETY

 CAUTION: Always wear safety goggles and a lab apron to protect your eyes and clothing.

CAUTION: Do not touch or taste any chemicals. Know the location of the emergency shower and eyewash station and how to use them. If you get a chemical on your skin or clothing, wash it off at the sink while calling to the teacher. Notify the teacher of a spill. Spills should be cleaned up promptly, according to your teacher's directions.

 CAUTION: Glassware is fragile. Notify the teacher of broken glass or cuts. Do not clean up broken glass or spills with broken glass unless the teacher tells you to do so.

Before You Begin

Sea stars are members of the phylum Echinodermata, a group of invertebrates that also includes sand dollars, sea urchins, and sea cucumbers. **Echinoderms** share four main characteristics: an **endoskeleton, five-part radial symmetry,** a **water-vascular system,** and circulation and respiration through their **coelom.**

1. Write a definition for each boldface term above. Use a separate sheet of paper.

2. You will be using the data table provided to record your data.

Analyzing Sea Star Anatomy *continued*

3. Based on the objectives for this lab, write a question you would like to explore about sea star anatomy.

_____

_____

Procedure

PART A: EXTERNAL ANATOMY

1. **CAUTION: Put on safety goggles, a lab apron, and protective gloves.** As you observe the sea star body structures, record your observations and your inference of each structure's function in the table. On a separate sheet of paper or in your lab notebook, draw and label the sea star and the structures that you observe.

Data Table		
Function of Sea Star Structures		
Structure	**Observations**	**Inferred function**
Madreporite		
Spine		
Skin gill		

2. Using forceps, hold a preserved sea star under running water to gently but thoroughly remove excess preservative. Then place the sea star in a dissecting tray.

3. Refer to a diagram of a sea star in your textbook to locate the madreporite on the upper surface of the sea star.

4. Use a hand lens to observe the sea star's spines. Are they distributed in any recognizable pattern? Are they exposed or covered by tissue? Are they movable or fixed?

_____

_____

5. Use the dissecting microscope to look for small skin gills, If any are present, describe their location and structure.

_____

_____

6. Examine the sea star's lower surface. Find the mouth, and use forceps or a probe to gently move aside any soft tissues. What structures are found around the mouth?

_____

7. Locate the tube feet. Describe their distribution. Using a dissecting microscope, observe and then draw a single tube foot on a separate sheet of paper.

_____

PART B: INTERNAL ANATOMY

8. **CAUTION: Scissors, probes, and pins are sharp. Use care not to puncture your gloves or injure yourself or others.** Using scissors and forceps, carefully cut the body wall away from the upper surface of one of the sea star's arms. Start near the end of the arm and work toward the center.

9. Find the digestive glands in the arm you have opened. Then, locate the short branched tube that connects the digestive glands to the pyloric stomach.

10. Cut the tube that connects the digestive glands to the stomach, and move the digestive glands out of the arm. Look for the reproductive organs.

11. Locate the two rows of ampullae that run the length of the arm.

12. Carefully remove the body wall from the upper surface of the central region of the sea star. Locate the pyloric stomach and the cardiac stomach.

13. Remove the stomachs and find the ring canal and the radial canals. In which direction does water move through these canals?

_____

14. Dispose of sea stars and sea star body parts in the waste container designated by your teacher. Do not put lab materials in the trash unless your teacher tells you to do so.

15. Clean up your work area and all lab equipment. Return lab equipment to its proper place. Wash your hands thoroughly before you leave the lab and after you finish all work.

Analyze and Conclude

1. Analyzing Results What type of symmetry is found in the sea star?

_____

2. Inferring Relationships What is the relationship between the ampullae and the tube feet?

_____

_____

_____

3. Making Predictions How does a sea star use its stomach during feeding?

_____

_____

_____

4. Making Predictions If the ring canals and radial canals did not function properly, how would this affect the sea star's ability to move and feed?

_____

_____

_____

5. Further Inquiry Write a new question about echinoderms that could be explored with another investigation.

_____

_____

_____

Name _____ Class _____ Date _____

Comparing Hominid Skulls

SKILLS
- Measuring
- Comparing anatomical features

OBJECTIVES
- **Identify** differences and similarities between the skulls of apes and the skulls of humans.
- **Identify** differences and similarities between the fossilized skulls of hominids.
- **Classify** the features of hominid skulls as apelike, humanlike, or intermediate.

MATERIALS
- metric ruler
- protractor

Before You Begin

Modern **apes** and humans share a **common ancestor.** Much of our understanding of human evolution is based on the study of the fossilized remains of **hominids.** By studying fossilized bones and identifying similar and dissimilar structures, scientists can infer the **anatomy,** or body structure, of a species. In this lab, you will identify differences and similarities between the skulls of apes, early hominids, and humans.

1. Write a definition for each boldface term in the paragraph above. Use a separate sheet of paper.

2. You will be using the data table provided to record your data.

3. Based on the objectives for this lab, write a question you would like to explore about human evolution.

_____

_____

_____

Comparing Hominid Skulls *continued*

How to Interpret the Features of a Skull

Cranial capacity: Use the circles drawn on the skulls to estimate brain volume, or cranial capacity. Measure the radius of each circle in centimeters. Then cube this number, and multiply the result by 1,000 to calculate the approximate life-size cranial capacity in cubic centimeters.

Lower face area: Measure A to B and C to D in centimeters for each skull. Multiply these two numbers together, and multiply the product by 40 to approximate the life-size lower face area in square centimeters.

Brain area: Measure E to F and G to H in centimeters for each skull. Multiply these two numbers and multiply the product by 40 to approximate the life-size brain area in square centimeters.

Jaw angle: Note the two lines that come together near the nose of each skull. Use a protractor to measure the inside angle made by the lines and to determine how far outward the jaw projects.

Brow ridge: Note the presence or absence of a bony ridge above the eye sockets.

Teeth: Count the number of each kind of teeth in the lower jaw.

Procedure

PART A: APE SKULLS AND HUMAN SKULLS

1. Examine the diagrams of the skull and jaw of an ape and a human. Look for similarities and differences between the features listed in the chart "How to Interpret the Features of a Skull." Record your observations and measurements for each feature listed in Data Table 1.

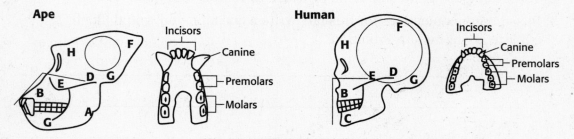

Data Table 1						
Name	Cranial capacity (cm³)	Lower face area (cm²)	Brain area (cm²)	Jaw angle (degrees)	Brow ridge	Teeth
Ape						
Human						

Comparing Hominid Skulls *continued*

PART B: FOSSIL HOMINIDS

2. Examine the four fossil hominid skulls. On the hominid skulls, observe and measure four features that are listed in the chart "How to Interpret the Features of a Skull." Use the human skull as a model for taking measurements. Record your observations and measurements in Data Table 2.

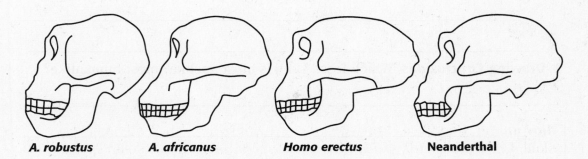

A. robustus   **A. africanus**   **Homo erectus**   **Neanderthal**

Data Table 2				
Name				
Australopithecus robustus				
Australopithecus africanus				
Homo erectus				
Neanderthal				

3. Compare your data for the hominids with your data for the modern ape and human. Classify each feature of the hominid skulls as being apelike, human-like, or intermediate by writing an *A*, *H*, or *I* next to your observation or measurement for that feature.

4. Using your data, predict the order in which the hominids shown here may have evolved.

_____

_____

Comparing Hominid Skulls *continued*

Analyze and Conclude

1. **Summarizing Results** How did skull structure change as hominids evolved?

_____

_____

_____

_____

2. **Drawing Conclusions** Which fossil skull is most apelike? most humanlike?

_____

_____

_____

_____

3. **Further Inquiry** Write a new question about human evolution that could be explored with another investigation.

_____

_____

_____

_____

Observing a Live Frog

SKILLS

- Observing
- Relating

OBJECTIVES

- **Examine** the external features of a frog.
- **Observe** the behavior of a frog.
- **Explain** how a frog is adapted to life on land and in water.

MATERIALS

- live frog in a terrarium
- live insects (crickets or mealworms)
- 600 mL beaker
- aquarium half-filled with dechlorinated water

Before You Begin

Frogs, which are **amphibians,** are adapted for living on land and in water. For example, a frog's eyes have an extra eyelid called the **nictitating membrane.** This eyelid protects the eye when the frog is underwater and keeps the eye moist when the frog is on land. The smooth skin of a frog acts as a respiratory organ by exchanging oxygen and carbon dioxide with the air or water. The limbs of a frog enable it to move both on land and in water. In this lab, you will examine a live frog in both a terrestrial environment and an aquatic environment.

1. Write a definition for each boldface term in the paragraph above and for the following term: tympanic membrane. Use a separate sheet of paper.

2. You will be using the data table provided to record your data.

3. Based on the objectives for this lab, write a question you would like to explore about frogs.

_____

_____

Procedure

PART A: OBSERVING A FROG

1. Observe a live frog in a terrarium. Closely examine the external features of the frog. On a separate sheet of paper, make a drawing of the frog. Label the eyes, nostrils, tympanic membranes, front legs, and hind legs.

Observing a Live Frog *continued*

Data Table	
Behavior/structure	**Observations**
Breathing	
Eyes	
Legs	
Response to food	
Response to noise	
Skin	
Swimming behavior	

2. Watch the frog's movements as it breathes air into and out of its lungs. Record your observations.

3. Look closely at the frog's eyes, and note their location. Examine the upper and lower eyelids as well as a third transparent eyelid called a *nictitating membrane*. Describe how the eyelids move.

4. Study the frog's legs, and note the difference between the front and hind legs.

5. Place a live insect, such as a cricket or a mealworm, into the terrarium. Observe how the frog reacts.

6. Tap the side of the terrarium farthest from the frog, and observe the frog's response.

7. Place a 600 mL beaker in the terrarium. **CAUTION: Handle live frogs gently. Frogs are slippery! Do not allow a frog to injure itself by jumping from a lab table to the floor.** Carefully pick up the frog, and examine its skin. How does it feel? Now place the frog in the beaker. Cover the beaker with your hand, and carry it to a freshwater aquarium. Tilt the beaker, and gently lower it into the water until the frog swims out.

8. Watch the frog float and swim. Notice how the frog uses its legs to swim. Also notice the position of the frog's head. As the frog swims, bend down to view the underside of the frog. Then look down on the frog from above. Compare the color on the dorsal and ventral sides of the frog.

PART B: CLEANUP AND DISPOSAL

9. Dispose of broken glass in the designated waste containers. Put live animals in the designated containers. Do not pour chemicals down the drain or put lab materials in the trash unless your teacher tells you to do so.

10. Clean up your work area and all lab equipment. Return lab equipment to its proper place. Wash your hands thoroughly before you leave the lab and after you finish all work.

Observing a Live Frog *continued*

Analyze and Conclude

1. **Summarizing Information** How does a frog use its hind legs for moving on land and in water?

_____

2. **Recognizing Relationships** How does the position of a frog's eyes benefit the frog while it is swimming?

_____

_____

3. **Analyzing Data** What features of an adult frog provide evidence that it has an aquatic life and a terrestrial life?

_____

_____

4. **Analyzing Methods** Were you able to determine in this lab how a frog hears? Explain.

_____

_____

_____

5. **Inferring Conclusions** What can you infer about a frog's field of vision from the position of its eyes?

_____

6. **Forming Hypotheses** How is the coloration on the dorsal and ventral sides of a frog an adaptive advantage?

_____

_____

_____

7. **Further Inquiry** Write a new question about frogs that could be explored with another investigation.

_____

_____

Observing Color Change in Anoles

SKILLS
- Using scientific methods
- Observing

OBJECTIVES
- **Observe** live anoles.
- **Relate** the color of an anole to the color of its surroundings.

MATERIALS
- glass-marking pencil
- 2 large, clear jars with wide mouths and lids with air holes
- 2 live anoles
- 6 shades each of brown and green construction paper, ranging from light to dark (2 swatches of each shade)

Before You Begin

Lizards are a group of **reptiles.** There are 250–300 species of anoles, lizards in the genus *Anolis*. Like chameleons, anoles can change color, ranging from brown to green. Anoles live in shrubs, grasses, and trees. Light level, temperature, and other factors, such as whether the animal is frightened or has eaten recently, can all affect the color of an anole. When anoles are frightened, they usually turn dark gray or brown and are unlikely to respond to other **stimuli.** Anoles generally change color within a few minutes. In this lab, you will observe the ability of anoles to change color when they are placed on different background colors. You will also determine how this ability might be an advantage to anoles.

1. Write a definition for each boldface term in the paragraph above.

_____

_____

_____

2. You will be using the data table provided to record your data.

3. Based on the objectives for this lab, write a question you would like to explore about the color-changing behavior of anoles.

_____

_____

Procedure

PART A: MAKE OBSERVATIONS

1. Observe live anoles in a terrarium. Make a list of characteristics that indicate that anoles are reptiles.

_____

_____

2. Work with a partner to place anoles to be studied in separate glass jars. **CAUTION: Handle anoles gently, and follow instructions carefully. Anoles run fast and are easily frightened. Plan your actions before you start.** By working efficiently, you can keep your anole from becoming overly frightened. Carefully pick up one anole by grasping it firmly but gently around the shoulders. Do not pick up anoles by their tail. Place the anole in a glass jar. Quickly and carefully place a lid with air holes on the jar.

3. When anoles become overly frightened, they remain dark. While you are designing your experiment, do not disturb your anoles, and let them recover from your handling.

PART B: DESIGN AN EXPERIMENT

4. Work with members of your lab group to explore one of the questions written for step 3 of **Before You Begin.** To explore the question, design an experiment that uses the materials listed for this lab.

> **You Choose**
>
> As you design your experiment, decide the following:
>
> **a.** what question you will explore
>
> **b.** what hypothesis you will test
>
> **c.** how many anoles you will need
>
> **d.** what background colors you will use
>
> **e.** how many times you will test each background with an anole
>
> **f.** how long you will observe each test and how you will keep track of time
>
> **g.** what your control will be
>
> **h.** what data to record in your data table

Observing Color Change in Anoles *continued*

5. Write a procedure for your experiment. Make a list of all the safety precautions you will take. Have your teacher approve your procedure and safety precautions before you begin the experiment.

Data Table

	Color 1		Color 2	
Anole	Change	Time	Change	Time
1				
2				

6. Set up and carry out your experiment.

PART C: CLEANUP AND DISPOSAL

7. Dispose of construction paper and broken glass in the designated waste containers. Put anoles in the designated container. Do not put lab materials in the trash unless your teacher tells you to do so.

8. Clean up your work area and all lab equipment. Return lab equipment to its proper place. Wash your hands thoroughly before you leave the lab.

Analyze and Conclude

1. Summarizing Results Briefly state how the variable you tested influenced the color-changing behavior of anoles.

_____

_____

_____

2. Evaluating Results Did any unplanned variables influence your data? (For example, was there a loud noise, or was a jar suddenly moved?)

_____

_____

_____

3. Analyzing Methods How could your experiment be modified to improve the certainty of your results?

_____

_____

4. Analyzing Data Were there any inconsistencies in your data? (For example, two anoles reacted in different ways.) If so, offer an explanation for them.

_____

_____

_____

5. Drawing Conclusions After considering your data, make a statement about color-changing behavior in anoles.

_____

_____

_____

6. Further Inquiry Write a new question about anoles that could be explored with another investigation.

_____

_____

_____

Exploration Lab

Exploring Mammalian Characteristics

SKILLS

- Observing
- Drawing
- Inferring

OBJECTIVES

- **Examine** distinguishing characteristics of mammals.
- **Infer** the functions of mammalian structures.

MATERIALS

- hand lens or stereomicroscope
- prepared slide of mammalian skin
- compound microscope
- mirror
- specimens or pictures of vertebrate skulls
  (some mammalian, some nonmammalian)

Before You Begin

Mammals are vertebrates with **hair, mammary glands,** a single lower jawbone, and specialized teeth. Other characteristics of mammals include **endothermy** and a four-chambered heart. Mammals also have **oil (sebaceous) glands** in their skin, and most have **sweat glands.** In this lab, you will examine some of the characteristics of mammals that distinguish them from other vertebrates.

1. Write a definition for each boldface term in the paragraph above. Use a separate sheet of paper.

2. Record your data in the data table provided.

3. Based on the objectives for this lab, write a question you would like to explore about the characteristics of mammals.

_____

_____

Procedure

PART A: EXAMINING MAMMALIAN SKIN

1. Use a hand lens to look at several areas of your skin, including areas that appear to be hairless. Record your observations.

_____

_____

2. Look at a prepared slide of mammalian skin under low power of a compound microscope. Notice the glands in the skin. Look for the oil (sebaceous) glands and the sweat glands., draw and label an example of each type of gland. Use a separate sheet of paper.

PART B: EXAMINING MAMMALIAN TEETH AND SKULLS

3. Wash your hands thoroughly with soap and water. Use a mirror to look in your mouth. Identify the four kinds of mammalian teeth you see.

_____

_____

4. Count each kind of tooth on one side of your lower jaw. Multiply the number of each kind of tooth by 4, and record these numbers in the appropriate columns of the data table below. Wash your hands again before continuing.

5. Look at the skulls of several mammals. Identify the kinds of teeth in each skull. For each skull, find the number of each kind of tooth as you did in step 4.

Data Table				
Mammal	**Incisors**	**Canines**	**Premolars**	**Molars**
Human				

6. Look at the skulls of several nonmammalian vertebrates, and compare nonmammalian teeth to mammalian teeth.

_____

_____

7. Compare the jaws of mammalian skulls to those of nonmammalian vertebrates. As you look at each skull, notice the structure of the lower jawbone and how the upper jawbone and the lower jawbone connect.

PART C: CLEANUP AND DISPOSAL

8. Dispose of broken glass in the waste container designated by your teacher.

9. Clean up your work area and all lab equipment. Return lab equipment to its proper place. Wash your hands thoroughly before you leave the lab and after you finish all work.

Analyze and Conclude

1. Summarizing Information List the characteristics that distinguish mammals from other vertebrates.

_____

_____

2. Interpreting Graphics Compare the amount of hair on humans to that on a skunk and a dolphin.

_____

_____

3. Inferring Relationships What role might hair or fur play in enabling mammals to be endotherms?

_____

_____

4. Forming Hypotheses Besides the role of hair you identified in item 3 above, what other roles do you think hair might play in mammals?

_____

_____

_____

5. Recognizing Patterns Where are the oil (sebaceous) glands located in the skin of mammals?

_____

Exploring Mammalian Characteristics *continued*

6. Forming Hypotheses Do you think skunks and dolphins have more sweat glands or fewer sweat glands than humans have? Explain.

_____

_____

_____

_____

7. Comparing Structures How is the mammalian jaw different from nonmammalian jaws?

_____

8. Inferring Conclusions Based on the shape of your teeth, would you classify humans as carnivores (meat eaters), herbivores (plant eaters), or omnivores (meat and plant eaters)? Explain.

_____

_____

_____

9. Evaluating Conclusions Justify the following conclusion: The kinds and shapes of a mammal's teeth can be used to determine its diet.

_____

_____

_____

10. Further Inquiry Write a new question about the characteristics of mammals that could be explored with a new investigation.

_____

_____

Studying Nonverbal Communication

SKILLS

- Observing
- Analyzing
- Graphing

OBJECTIVES

- **Recognize** that posture is a type of nonverbal communication.
- **Observe** how human posture changes during a conversation.
- **Determine** the relationship of gender to the postural changes that occur during a conversation.

MATERIALS

- stopwatch or clock with a second hand
- paper
- pencil

Before You Begin

People communicate nonverbally with their **posture**, or body position. The position of the body while standing is called the **stance.** In an **equal stance,** the body weight is supported equally by both legs. In an **unequal stance,** more weight is supported by one leg than by the other. In this lab, you will observe and analyze how stance changes during conversations between pairs of people who are standing.

1. Write a definition for each boldface term in the paragraph above. Use a separate sheet of paper.

2. You will be using the data table provided to record your data. Notice that the second and third tables are continuations of the first. Use a separate sheet of paper if you need to make new data tables for longer conversations.

3. Based on the objectives for this lab, write a question you would like to explore about nonverbal communication.

_____

_____

Studying Nonverbal Communication *continued*

Procedure
PART A: OBSERVING BEHAVIOR

1. Work in a group of two or three to observe conversations between pairs of people. Each conversation must last between 45 seconds and 5 minutes. One person in your group should be the timekeeper and the other group members should record data. **Note:** Be sure that your subjects are unaware they are being observed.

Data Table					
Pairs	**Gender Involved**	**Gender Observed**	**15 second intervals**		
			15 s	**30 s**	**45 s**
1					
2					
3					

Data Table					
Pairs	**Gender Involved**	**Gender Observed**	**15 second intervals**		
			1 min	**1 min 15 s**	**1 min 30 s**
1					
2					
3					

Data Table					
Pairs	**Gender Involved**	**Gender Observed**	**15 second intervals**		
			1 min 45 s	**2 min**	**2 min 15 s**
1					
2					
3					

2. Observe at least three conversations. Record the genders of the two participants in each conversation and the gender of the one person whose posture you observe. **Note:** Be sure that the timekeeper accurately clocks the passage of each 15-second interval.

3. For each 15-second interval, record all of the changes in stance by the person you are observing. For example, note every time your subject shifts from an equal stance to an unequal stance, or vice versa. To record the stance simply, you may write *E* to identify an equal stance and *U* to identify an unequal stance.

4. If the subject assumes an unequal stance, also record the number of weight shifts from one foot to the other. Indicate a weight shift simply by writing *W*.

5. When a conversation ends, write down whether the pair departed together or separately. To record this, write *T* to indicate departing together or *S* to indicate departing separately.

6. After you have completed each observation, tally the total number of weight shifts within each 15-second block. **IMPORTANT!** Retain data only for conversations that last at least 45 seconds.

If a conversation ends before you have collected data for 45 seconds, observe another conversation.

PART B: ANALYZING DATA

7. After all observations have been completed, combine the data from all of the groups in your class. Analyze the data, without regard to gender.

   a. Determine the most common stance during the first 15 seconds of a conversation, the middle 15 seconds, and the last 15 seconds. Make a bar graph on a separate sheet of paper to summarize the class data.

   _____

   _____

   _____

   b. Find the average number of weight shifts in the beginning, middle, and end intervals. Make a bar graph on a separate sheet of paper to summarize the class data.

   _____

   _____

   _____

8. Repeat step 7, but analyze the data according to gender this time.

   _____

   _____

9. Compile the data and make bar graphs on a separate sheet of paper for each of the following: males talking with a male, males talking with a female, females talking with a male, and females talking with a female. Compare these graphs with the ones you made in step 7.

   _____

   _____

Studying Nonverbal Communication *continued*

Analyze and Conclude

1. **Analyzing Results** Which stance was used most often during a conversation?

_____

_____

2. **Recognizing Relationships** Which behavior most often signals that a conversation is about to end: stance change or weight shift?

_____

_____

3. **Drawing Conclusions** Do males and females differ in their departure signals? Justify your conclusion.

_____

_____

_____

4. **Forming Hypotheses** What do you think might be an adaptive significance of a departure signal?

_____

_____

_____

5. **Forming Reasoned Opinions** What other behaviors you observed were forms of nonverbal communication? Justify your answer.

_____

_____

6. **Further Inquiry** Write a new question about animal behavior that could be explored with a new investigation.

_____

_____

_____

Name _____ Class _____ Date _____

Analyzing the Work of Muscles

SKILLS

- Using scientific methods
- Data collection
- Data interpretation

OBJECTIVE

- **Relate** muscles to the work they do.
- **Observe** the effects of fatigue

MATERIALS

- watch with second hand
- graph paper
- spring hand grips

Before You Begin

Muscles are attached to bones. As muscles contract, they move the bones to which they are attached. This is a basic type of work accomplished by the human body. As muscles are used, lactic acid builds up, resulting in **fatigue.** In this lab you will investigate how fatigue affects the amount of work that muscles can do.

1. Write a definition for the boldface term in the preceding paragraph.

_____

2. Use the data table below.

Data Table For Muscle Contractions													
Number of Muscle Contractions in 10-Second Intervals													
Time	1st 10 sec	2nd 10 sec	3rd 10 sec	4th 10 sec	5th 10 sec	6th 10 sec	7th 10 sec	8th 10 sec	9th 10 sec	10th 10 sec	11th 10 sec	12th 10 sec	13th 10 sec	14th 10 sec
Trial 1														
Trial 2														
Trial 3														
Trial 4														

PROCEDURE

1. Perform this investigation with a partner. Designate one laboratory partner to observe and record while the other performs the experiment.

2. Hold the spring hand grips in your left hand if you are right-handed, or in your right hand if you are left-handed. Squeeze the grips rapidly and as hard as possible at a steady pace, until complete fatigue is experienced in the muscles of your hand and forearm.

3. The recorder should count and record the number of squeezes for every 10 seconds.

4. Allow the experimenter to rest for 1 minute and repeat the procedure for two more trials.

5. Record the data in your table. Some spaces may be left blank.

6. Switch roles with your partner and repeat the procedure.

Analyze and Conclude

1. **Summarizing Results** Use graph paper to plot the results of the three trials on a graph. The x-axis should be used for time in seconds and the y-axis for the number of muscle contractions.

2. **Analyzing Data** Account for the differences in the amount of work done by the muscles during the three trials.

   _____

   _____

3. **Drawing Conclusions** What is the relationship between the work muscles can do and fatigue?

   _____

   _____

   _____

4. **Predicting Patterns** How does the work done in the muscles of your hands and arms relate to the work done by the muscle of your heart?

   _____

   _____

5. **Further Inquiry** Compare the charts and the graphs of the athletes and nonathletes.

   _____

   _____

Determining Lung Capacity

SKILLS

- Measuring
- Organizing data
- Comparing

OBJECTIVES

- **Measure** your tidal volume, vital capacity, and expiratory reserve volume.
- **Determine** your inspiratory reserve capacity and lung capacity.
- **Predict** how exercise will affect tidal volume, vital capacity, and lung capacity.

MATERIALS

- spirometer
- spirometer mouthpiece

Before You Begin

Lung capacity is the total volume of air that the lungs can hold. The lung capacity of an individual is influenced by many factors, such as gender, age, strength of diaphragm and chest muscles, and disease.

During normal breathing, only a small percentage of your lung capacity is inhaled and exhaled. The amount of air inhaled or exhaled in a normal breath is called the **tidal volume.** An additional amount of air, called the **inspiratory reserve volume,** can be forcefully inhaled after a normal inhalation. The **expiratory reserve volume** is the amount of air that can be forcefully exhaled after a normal exhalation. **Vital capacity** is the maximum amount of air that can be inhaled or exhaled. Even after you have exhaled all the air you can, a significant amount of air called the **residual volume** still remains in your lungs.

In this lab, you will determine your lung capacity by using a **spirometer,** which is an instrument used to measure the volume of air exhaled from the lungs.

1. Write a definition for each boldface term in the paragraph above. Use a separate sheet of paper.

2. You will be using the data table provided to record your data.

3. Based on the objectives for this lab, write a question about breathing that you would like to explore.

_____

_____

Determining Lung Capacity *continued*

Procedure

PART A: MEASURING VOLUME

1. Place a clean mouthpiece in the end of a spirometer. **CAUTION: Many diseases are spread by body fluids, such as saliva. Do NOT share a spirometer mouthpiece with anyone.**

2. To measure your tidal volume, first inhale a normal breath. Then exhale a normal breath into the spirometer through the mouthpiece. Record the volume of air exhaled in the data table below.

Data Table	
Tidal volume	
Expiratory reserve volume	
Inspiratory reserve volume	
Vital capacity	
Estimated residual volume	
Estimated lung capacity	

3. To measure your expiratory reserve volume, first inhale a normal breath and then exhale normally. Then forcefully exhale as much air as possible into the spirometer. Record this volume.

4. To measure your vital capacity, first inhale as much air as you can, and then forcefully exhale as much air as you can into the spirometer. Record this volume.

PART B: CALCULATING LUNG CAPACITY

The table below contains average values for residual volumes and lung capacities for young adults.

Residual Volumes and Lung Capacities	Males	Females
Residual volume*	1,200 mL	900 mL
Lung capacity*	6,000 mL	4,500 mL

*Athletes can have volumes 30–40% greater than the average for their gender.

5. Inspiratory reserve volume (IRV) can be calculated by subtracting tidal volume (TV) and expiratory reserve volume (ERV) from vital capacity (VC). The formula for this calculation is as follows:

$$IRV = VC - TV - ERV$$

Use the data in the data table and the equation above to calculate your estimated inspiratory reserve volume.

Determining Lung Capacity *continued*

6. Lung capacity (LC) can be calculated by adding residual volume (RV) to vital capacity (VC). The formula for this calculation is as follows:

$$LC = VC + RV$$

Use the data in the data table and the table above to calculate your estimated lung capacity.

PART C: CLEANUP AND DISPOSAL

7. Dispose of your mouthpiece in the designated waste container.

8. Clean up your work area and all lab equipment. Return lab equipment to its proper place. Wash your hands thoroughly before you leave the lab and after you finish all work.

Analyze and Conclude

1. Interpreting Data How does your expiratory reserve volume compare with your inspiratory reserve volume?

_____

_____

2. Interpreting Tables How does the residual volume and lung capacity of an average young adult female compare with those of an average young adult male?

_____

_____

_____

3. Analyzing Data How did your tidal volume compare with that of others?

_____

_____

4. Recognizing Relationships Why was the value you found for your lung capacity an estimated value?

_____

_____

_____

_____

_____

_____

5. Analyzing Methods Why didn't you measure inspiratory reserve volume directly?

_____

_____

_____

_____

_____

6. Inferring Conclusions Why would males and athletes have greater vital capacities than females?

_____

_____

_____

_____

_____

7. Justifying Conclusions Use data from your class to justify the conclusion that exercise increases lung capacity.

_____

_____

_____

8. Further Inquiry Write a new question that could be explored with another investigation.

_____

_____

Name _____ Class _____ Date _____

Exploration Lab

Demonstrating Lactose Digestion

SKILLS

- Using scientific methods
- Observing
- Comparing

OBJECTIVES

- **Describe** the relationship between enzymes and the digestion of food molecules.
- **Evaluate** the ability of a milk-treatment product to promote lactose digestion.
- **Infer** the presence of lactose in milk and foods that contain milk.

MATERIALS

- milk-treatment product (liquid)
- toothpicks
- depression slides
- droppers
- whole milk
- glucose solution
- glucose test strips

SAFETY

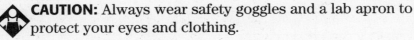 **CAUTION:** Always wear safety goggles and a lab apron to protect your eyes and clothing.

 CAUTION: Do not touch or taste any chemicals. Know the location of the emergency shower and eyewash station and how to use them. If you get a chemical on your skin or clothing, wash it off at the sink while calling to the teacher. Notify the teacher of a spill. Spills should be cleaned up promptly, according to your teacher's directions.

 CAUTION: Glassware is fragile. Notify the teacher of broken glass or cuts. Do not clean up broken glass or spills with broken glass unless the teacher tells you to do so.

Before You Begin

People with a condition known as **lactose intolerance** often experience stomach and intestinal pain, bloating, and diarrhea when they eat foods that contain milk. These symptoms result from an inability to digest lactose, a sugar found in milk. **Lactose** is a disaccharide made of one glucose unit and one galactose unit. Lactose molecules are broken down into glucose and galactose molecules during **digestion.** People who cannot digest lactose do not produce **lactase,** the digestive enzyme that aids the breakdown of lactose. In this lab, you will investigate a milk-treatment product that is designed to aid lactose digestion.

1. Write a definition for each boldface term in the paragraph above. Use a separate sheet of paper.

2. List at least 10 foods that contain milk.

_____

_____

3. You will be using the data table provided to record your data.

4. Based on the objectives for this lab, write a question you would like to explore about enzymes and digestion.

_____

_____

Data Table		
Solution	**Result (+ or −)**	**Interpretation**

Demonstrating Lactose Digestion *continued*

Procedure

PART A: DESIGN AN EXPERIMENT

1. Read the information sheet that comes with the milk-treatment product. Discuss with your lab group what the product is and what it does. Write a summary of your discussion for your lab report. Use a separate sheet of paper.

2. Work with the members of your lab group to explore one of the questions written for step 4 of **Before You Begin.** To explore the question, design an experiment that uses the materials listed for this lab.

> **You Choose**
>
> As you design your experiment, decide the following:
>
>   **a.** what question you will explore
>
>   **b.** what hypothesis you will test
>
>   **c.** what your controls will be
>
>   **d.** how much milk and milk-treatment product to use for each test
>
>   **e.** how to determine whether lactose was broken down
>
>   **f.** what data to record in your data table

3. Write the procedure for your group's experiment. Make a list of all the safety precautions you will take. Have your teacher approve your procedure and safety precautions before you begin the experiment.

4. ◆ ◆ ◆ ◆ Set up your group's experiment, and collect data.

PART B: CLEANUP AND DISPOSAL

5. ◆ Dispose of solutions, broken glass, and glucose test strips in the designated waste containers. Do not pour chemicals down the drain or put lab materials in the trash unless your teacher tells you to do so.

6. ◆ Clean up your work area and all lab equipment. Return lab equipment to its proper place. Wash your hands thoroughly before you leave the lab and after you finish all work.

Analyze and Conclude

1. **Summarizing Information** What are the milk-treatment product's ingredients?

_____

_____

Demonstrating Lactose Digestion *continued*

2. Recognizing Relationships What is the relationship between lactose and lactase?

_____

_____

_____

3. Analyzing Methods What role did the glucose solution play in your experiment?

_____

4. Drawing Conclusions What does the milk-treatment product do to milk?

_____

5. Analyzing Conclusions How do your results justify your conclusion?

_____

_____

_____

6. Evaluating Methods Why should you test the milk-treatment product with glucose test strips?

_____

7. Analyzing Results What do you infer from the results of this lab about treatments for other medical problems resulting from enzyme deficiencies?

_____

_____

8. Forming Reasoned Opinions As a person grows older, will he or she be more likely or less likely to develop lactose intolerance? Explain your answer.

_____

_____

_____

9. Predicting Patterns Do you think lactose intolerance might be inherited? Explain your answer.

_____

_____

_____

10. Further Inquiry Write a new question about enzymes and digestion that could be explored with another investigation.

_____

_____

_____

Simulating Disease Transmission

SKILLS

• Modeling

• Organizing and analyzing data

OBJECTIVES

• **Simulate** the transmission of a disease.

• **Determine** the original carrier of the disease.

MATERIALS

• safety goggles

• lab apron

• disposable gloves

• dropper bottle of unknown solution

• large test tube

• indophenol indicator

SAFETY

 CAUTION: Always wear safety goggles and a lab apron to protect your eyes and clothing.

 CAUTION: Do not touch or taste any chemicals. Know the location of the emergency shower and eyewash station and how to use them. If you get a chemical on your skin or clothing, wash it off at the sink while calling to the teacher. Notify the teacher of a spill. Spills should be cleaned up promptly, according to your teacher's directions.

 CAUTION: Glassware is fragile. Notify the teacher of broken glass or cuts. Do not clean up broken glass or spills with broken glass unless the teacher tells you to do so.

Before You Begin

Communicable diseases are caused by **pathogens** and can be transmitted from one person to another. You can become infected by a pathogen in several ways, including by drinking contaminated water, eating contaminated foods, receiving contaminated blood, and inhaling infectious **aerosols** (droplets from coughs or sneezes). In this lab, you will simulate the transmission of a communicable disease. After the simulation, you will try to identify the original infected person in the closed class population.

1. Write a definition for each boldface term in the paragraph above. Use a separate sheet of paper.

2. You will be using the data table provided to record your data.

3. Based on the objectives for this lab, write a question you would like to explore about disease transmission.

_____

_____

_____

Procedure

PART A: SIMULATE DISEASE TRANSMISSION

1. Put on safety goggles, a lab apron, and gloves.

2. You will be given a dropper bottle of an unknown solution. When your teacher says to begin, transfer 3 dropperfuls of your solution to a clean test tube.

3. Select a partner for Round 1. Record the name of this partner in Data Table 1.

Data Table 1	
Round number	**Partner's name**

4. Pour the contents of one of your test tubes into the other test tube. Then pour half the solution back into the first test tube. You and your partner now share any pathogens either of you might have.

5. On your teacher's signal, select a new partner for Round 2. Record this partner's name in Data Table 1. Repeat step 4.

6. On your teacher's signal, select another new partner for Round 3. Record this partner's name. Repeat step 4.

7. Add one dropperful of indophenol indicator to your test tube. "Infected" solutions will stay colorless or turn light pink. "Uninfected" solutions will turn blue. Record the results of your test.

Simulating Disease Transmission *continued*

PART B: TRACE THE DISEASE SOURCE

8. If you are infected, write your name and the name of your partner in each round on the board or on an overhead projector. Mark your infected partners. Record all the data for your class in Data Table 2.

Data Table 2			
Name of infected person	**Names of infected person's partners**		
	Round 1	**Round 2**	**Round 3**

9. To trace the source of the infection, cross out the names of the uninfected partners in Round 1. There should be only two names left. One is the name of the original disease carrier. To find the original disease carrier, place a sample from his or her dropper bottle in a clean test tube, and test it with indophenol indicator.

Simulating Disease Transmission *continued*

10. To show the disease transmission route, make a diagram similar to the one that follows. Show the original disease carrier and the people each disease carrier infected.

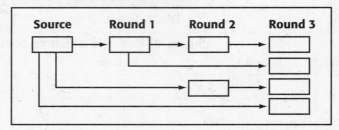

Disease Transmission Route

PART C: CLEANUP AND DISPOSAL

11. Dispose of solutions and broken glass in the designated waste containers. Do not pour chemicals down the drain unless your teacher tells you to do so.

12. Clean up your work area and all lab equipment. Return lab equipment to its proper place. Wash your hands thoroughly before you leave the lab and after you finish all work.

Analyze and Conclude

1. Interpreting Data After Round 3, how many people were "infected"? Express this number as a percentage of your class.

_____

2. Relating Concepts What do you think the clear fluids each student started with represent? Explain why.

_____

_____

3. Drawing Conclusions Can someone who does not show any symptoms of a disease transmit that disease? Explain.

_____

_____

4. Further Inquiry Write a new question about disease transmission that could be explored with another investigation.

_____

_____

Name _____ Class _____ Date _____

Calculating Reaction Times

SKILLS

- Measuring
- Calculating

OBJECTIVES

- **Determine** human reaction times.
- **Design** an experiment that measures changes in reaction times.

MATERIALS

- meterstick

Before You Begin

When you want to move your hand, your brain must send a message all the way to the muscles in your arms. How long does that take? In this exercise, you will work with a partner to see how quickly you can react. In this lab, you will investigate reaction times and design an experiment to investigate influences on reaction times.

1. You will be using the data table provided to record your data.

2. Write a hypothesis about an influence on reaction times. (For example: "People who have eaten breakfast have faster reaction times than people who have not eaten breakfast.")

_____

_____

_____

Procedure

PART A: CALCULATING REACTION TIMES

1. Sit in a chair and have a partner stand facing you while holding a meterstick in a vertical position.

2. Hold your thumb about 3 cm from your fingers near the bottom end of the stick. The meterstick should be positioned to fall between your thumb and fingers.

3. Tell your partner to let go of the meterstick without warning.

4. When your partner releases the meterstick, catch the stick by pressing your thumb and fingers together. Your partner should be ready to catch the top of the meterstick if it begins to tip over.

| Calculating Reaction Times *continued*

5. Record the number of centimeters the stick dropped before you caught it. The distance that the meterstick falls before you catch it can be used to evaluate your reaction time.

6. Repeat the procedure several times, and calculate the average number of centimeters.

7. Try this procedure with your other hand.

8. Close your eyes and have your partner say "now," when the stick is released.

9. Exchange places with your partner, and repeat the procedure.

Data Table		
Hand: trial number	**Subject 1 reaction time (s)**	**Subject 2 reaction time (s)**
Left: 1		
Left: 2		
Left: 3		
Left: average		
Right: 1		
Right: 2		
Right: 3		
Right: average		

PART B: DESIGNING YOUR OWN EXPERIMENT

10. Work with the members of your lab group to explore one of the hypotheses written in the **Before You Begin** section of this lab.

You Choose

As you design your experiment, decide the following:

a. what hypothesis you will explore

c. how you will test the hypothesis

d. what the controls will be

e. how many trials to perform

f. what data to record in your data table

11. Write a procedure for your experiment. Make a list of all the safety precautions you will take. Have your teacher approve your procedure and safety precautions before you begin the experiment.

12. Set up your group's experiment and collect data.

Calculating Reaction Times *continued*

Analyze and Conclude

1. Summarizing Results What was your fastest reaction time?

_____

2. Analyzing Data How does your reaction time when using your dominant hand compare with your reaction time when using your other hand?

_____

_____

3. Drawing Conclusions Why may each hand have a different reaction time? Why may each person have a different reaction time? Compared to earlier trials, was the reaction time in step 8 faster, slower or the same? If the time was faster or slower, hypothesize a reason for the difference.

_____

_____

_____

4. Predicting Patterns Compile the data gathered by each pair in your class. Can you identify any trends in the data? (For example, do males and females have the same average reaction times?)

_____

_____

_____

5. Further Inquiry Write a new question about reaction times that could be explored in another investigation.

_____

_____

_____

Exploration Lab

The Effect of Epinephrine on Heart Rate

SKILLS

- Using scientific methods
- Graphing
- Calculating

OBJECTIVES

- **Determine** the heart rate of *Daphnia*.
- **Observe** the effect of the hormone epinephrine on heart rate in *Daphnia*.
- **Determine** the threshold concentration for the action of epinephrine on *Daphnia*.

MATERIALS

- medicine droppers
- *Daphnia*
- *Daphnia* culture water
- depression slides
- petroleum jelly
- coverslips
- compound microscope
- watch or clock with second hand
- paper towels
- 100 mL beaker
- 10 mL graduated cylinders
- epinephrine solutions (0.001%, 0.0001%, 0.00001%, and 0.000001%)

SAFETY

 CAUTION: Always wear safety goggles and a lab apron to protect your eyes and clothing.

 CAUTION: Do not touch or taste any chemicals. Know the location of the emergency shower and eyewash station and how to use them. If you get a chemical on your skin or clothing, wash it off at the sink while calling to the teacher. Notify the teacher of a spill. Spills should be cleaned up promptly, according to your teacher's directions.

 CAUTION: Glassware is fragile. Notify the teacher of broken glass or cuts. Do not clean up broken glass or spills with broken glass unless the teacher tells you to do so.

| The Effect of Epinephrine on Heart Rate *continued*

Before You Begin

Epinephrine is a hormone released in response to stress. It increases blood pressure, blood glucose level, and **heart rate** (HR). The lowest concentration that stimulates a response is called the **threshold concentration.** In this lab, you will observe the effect of epinephrine on HR using the crustacean *Daphnia.* Epinephrine affects the HR of *Daphnia* and humans in similar ways.

1. Write a definition for each boldface term in the paragraph above. Use a separate sheet of paper.

2. You will be using the data table provided to record your data.

3. Based on the objectives for this lab, write a question you would like to explore about the action of hormones.

_____

_____

Procedure

PART A: OBSERVING HEART RATE IN *DAPHNIA*

1. **Caution: Do not touch your face while handling microorganisms.** Use a clean medicine dropper to transfer one *Daphnia* to the well of a clean depression slide. Place a dab of petroleum jelly in the well. Add a coverslip. Observe with a compound microscope under low power.

2. Count the *Daphnia*'s heartbeats for 10 seconds. Divide this number by 10 to find the HR in beats/s. Record this number under Trial 1 in your data table. Turn off the microscope light, and wait 20 seconds. Repeat the count for Trials 2 and 3.

Data Table					
Solution	**HR (beats/s) Trial 1 (A)**	**HR (beats/s) Trial 2 (B)**	**HR (beats/s) Trial 3 (C)**	**Average HR (beats/s) [(A+B+C)/3]**	**Average HR (beats/min)**

The Effect of Epinephrine on Heart Rate *continued*

3. After calculating the average HR in beats/s, calculate the HR in beats/min by using the following formula: HR (in beats/min) = Average HR (in beats/s) × 60 s/min.

PART B: DESIGN AN EXPERIMENT

4. Work with the members of your lab group to explore one of the questions written for step 3 of **Before You Begin.** To explore the question, design an experiment that uses the materials listed for this lab.

> **You Choose**
>
> As you design your experiment, decide the following:
>
> **a.** what question you will explore
>
> **b.** what hypothesis you will test
>
> **c.** how many *Daphnia* to use
>
> **d.** what your controls will be
>
> **e.** what concentrations of epinephrine to test
>
> **f.** how many trials to perform
>
> **g.** what data to record in your data table

5. Write a procedure for your experiment. Make a list of all the safety precautions you will take. Have your teacher approve your procedure and safety precautions before you begin the experiment.

PART C: CONDUCT YOUR EXPERIMENT

6. Put on safety goggles, gloves, and a lab apron.

7. To add a solution to a prepared slide, first place a drop of the solution at the edge of the coverslip. Then place a piece of paper towel along the opposite edge to draw the solution under the coverslip. Wait 1 minute for the solution to take effect.

8. Set up your group's experiment, and collect data. **Caution: Epinephrine is toxic and is absorbed through the skin.**

PART D: CLEANUP AND DISPOSAL

9. Dispose of solutions and broken glass in the designated waste containers. Place treated *Daphnia* in a "recovery container." Do not pour chemicals down the drain or put lab materials in the trash unless your teacher tells you to do so.

10. Clean up your work area and all lab equipment. Return lab equipment to its proper place. Wash your hands thoroughly before you leave the lab and after you finish all work.

The Effect of Epinephrine on Heart Rate *continued*

Analyze and Conclude

1. **Summarizing Results** Use graph paper to make a graph of your group's data. Plot "Epinephrine concentration (%)" on the x-axis. Plot "Average heart rate (beats/min)" on the y-axis.

2. **Analyzing Data** Which solutions affected the heart rate of *Daphnia*?

_____

3. **Drawing Conclusions** What was the threshold concentration of epinephrine?

_____

4. **Predicting Patterns** Based on the information you have and on your data, predict how epinephrine concentration would affect human heart rates.

_____

_____

5. **Further Inquiry** Write a new question about hormones that could be explored with another investigation.

_____

_____

Skills Practice Lab

Observing Embryonic Development

SKILLS

- Observing
- Comparing and contrasting
- Making drawings
- Drawing conclusions

OBJECTIVES

- **Identify** the stages of early animal development.
- **Describe** the changes that occur during early development.
- **Compare** the stages of human embryonic development with those of echinoderm embryonic development.

MATERIALS

- prepared slides of sea star development, including
  - unfertilized egg
  - zygote
  - 2-cell stage
  - 4-cell stage
  - 8-cell stage
  - 16-cell stage
  - 32-cell stage
  - 64-cell stage
  - blastula
  - early gastrula
  - middle gastrula
  - late gastrula
- compound light microscope
- paper and pencil

SAFETY

 CAUTION: Always wear safety goggles and a lab apron to protect your eyes and clothing.

 CAUTION: Glassware is fragile. Notify the teacher of broken glass or cuts. Do not clean up broken glass or spills with broken glass unless the teacher tells you to do so.

Observing Embryonic Development *continued*

Before You Begin

Most members of the animal kingdom begin life as a single cell—the fertilized egg, or **zygote.** The early stages of development are quite similar in different species. Cleavage follows fertilization. During cleavage, the zygote divides many times without growing. The new cells migrate and form a hollow ball of cells called a **blastula.** The cells then begin to organize into the three primary germ layers: endoderm, mesoderm, and ectoderm. During this process, the developing organism is called a **gastrula.**

1. Write a definition for each boldface term in the preceding paragraph. Use a separate sheet of paper.

_____

_____

_____

2. Based on the objectives for this lab, write a question you would like to explore about embryonic development.

_____

_____

Procedure

1. Obtain a set of prepared slides that show star eggs at different stages of development. Choose slides labeled unfertilized egg, zygote, 2-cell stage, 4-cell stage, 8-cell stage, 16-cell stage, 32-cell stage, 64-cell stage, blastula, early gastrula, middle gastrula, late gastrula, and young sea star larva. (Note: *Blastula* is the general term for the embryonic stage that results from cleavage. In mammals, a blastocyst is a modified form of the blastula.)

2. Examine each slide using a compound light microscope. Using the microscope's low-power objective first, focus on one good example of the developmental stage listed on the slide's label. Then switch to the high-power objective, and focus on the image with the fine adjustment.

3. In your lab report, draw a diagram of each developmental stage that you examine (in chronological order). Label each diagram with the name of the stage it represents and the magnification used. Record your observations as soon as they are made. Do not redraw your diagrams. Draw only what you see; lab drawings do not need to be artistic or elaborate. They should be well organized and include specific details.

4. Compare your diagrams with the diagrams of human embryonic stages shown on the next page.

| Observing Embryonic Development *continued*

2-cell stage   **4-cell stage**

8-cell stage   **64-cell stage**

Blastocyst

5. Clean up your materials and wash your hands before leaving the lab.

Analyze and Conclude

1. Summarizing Results Compare the size of the sea star zygote with that of the blastula. At what stage does the embryo become larger than the zygote?

_____

_____

2. Analyzing Data What is the earliest stage in which all of the cells in the embryo no longer look exactly alike? How do cell shape and size change during successive stages of development?

_____

_____

_____

_____

_____

Observing Embryonic Development *continued*

3. Drawing Conclusions From your observations of changes in cellular organization, why do you think the blastocoel (the space in the center of the hollow sphere of cells of a blastula) is important during embryonic development?

_____

_____

4. Predicting Patterns How are the symmetries of a sea star embryo and a sea star larva different from the symmetry of an adult sea star? Would you expect to see a similar change in human development? What must happen to the sea star gastrula before it becomes a mature sea star?

_____

_____

_____

_____

_____

5. Further Inquiry How do your drawings of sea star embryonic development compare with those of human embryonic development? Based on your observations, in what ways do you think sea star embryos could be used to study early human development?

_____

_____

_____

_____

_____

Exploration Lab

Observing the Effects of Acid Rain on Seeds

Teacher Notes

TIME REQUIRED Day 1: 45–55 minutes; Days 3–10: 15 minutes, every other day.

SKILLS ACQUIRED
Collecting Data
Designing Experiments
Analyzing Data
Predicting

RATING
Easy ◄——— 1 —— 2 —— 3 —— 4 ———► Hard

Teacher Prep–3
Student Setup–2
Concept Level–3
Cleanup–2

SCIENTIFIC METHODS

In this lab, students will:
Make Observations
Test the Hypothesis
Analyze the Results
Draw Conclusions

MATERIALS

Materials for this lab can be ordered from WARD'S. Use the Lab Materials QuickList Software on the **One-Stop Planner CD-ROM** for catalog numbers and to create a customized list of materials for this lab.

SAFETY CAUTIONS

Prepare solutions under a ventilated hood. Wear goggles, impermeable gloves, and a lab apron.

TIPS AND TRICKS

Prepare 600 mL of mold inhibitor for 25 students (one part concentrated bleach to four parts water). Prepare solutions of different pH as follows. Use distilled water to dilute 5 mL of 1.0 M sulfuric acid (H_2SO_4) to 1 L to prepare a 0.01 M H_2SO_4 solution with a pH of 2. Dilute 50 mL of the 0.01 M H_2SO_4 solution to 1 L to make a solution with a pH of 3. Repeat this procedure using 5 mL and 0.5 mL of the 0.01 M H_2SO_4 solution to make pH 4 and pH 5 solutions, respectively. Verify the pH of each solution. Allow two days between the start of the experiment and the first observation.

SAMPLE PROCEDURE

1. Label plastic bags "pH 3," "pH 4," "pH 5," and "control."

2. Moisten three layers of paper towels with each solution.

3. Arrange 10 seeds that have been treated with mold inhibitor on one half of each set of treated paper towels. Fold the other half over the seeds. Place paper towels and seeds in the proper bag and seal the bag.

4. Record the number of seeds germinated and the length of each seedling. Note any other changes in the seedlings.

5. After each observation, re-wet the paper towels with the same solution as noted on each bag. Return the seeds to the bag.

ANSWERS TO BEFORE YOU BEGIN

1. *acid rain*—rain with a pH lower than 7 due to the pollutants it contains

   experiment—a planned procedure to test a hypothesis

   observation—the act of noting or perceiving objects or events using the senses

   pH—a relative measure of the hydrogen ion concentration within a solution

   hypothesis—an explanation that might be true and that can be tested by additional observations or experimentation

   prediction—the expected outcome of a test, assuming the hypothesis is correct

   variable—factor that can change

   control group—a group in an experiment that receives no experimental treatment and serves as a standard with which experimental groups can be compared

Name _____ Class _____ Date _____

Observing the Effects of Acid Rain on Seeds

SKILLS

- Using scientific methods
- Collecting, organizing, and graphing data

OBJECTIVES

- **Use** a scientific method to investigate a problem.
- **Predict** how acid rain affects germination and growth.

MATERIALS

- safety goggles
- protective gloves
- lab apron
- 50 seeds
- 250 mL beakers
- 20 mL mold inhibitor
- distilled water
- paper towels
- solutions of different pH
- wax pencil or marker
- resealable plastic bags
- metric ruler
- graph paper

SAFETY

  CAUTION: Always wear safety goggles and a lab apron to protect your eyes and clothing.

 CAUTION: Do not touch or taste any chemicals. Know the location of the emergency shower and eyewash station and how to use them. If you get a chemical on your skin or clothing, wash it off at the sink while calling to the teacher. Notify the teacher of a spill. Spills should be cleaned up promptly, according to your teacher's directions.

CAUTION: Glassware is fragile. Notify the teacher of broken glass or cuts. Do not clean up broken glass or spills with broken glass unless the teacher tells you to do so.

Before You Begin

Living things, such as salamander embryos, can be damaged by **acid rain** at certain times during their lives. In this lab, you will investigate the effect of acidic solutions on seeds. One way to investigate a problem is to design and conduct an **experiment.** We begin a scientific investigation by making **observations** and asking questions.

1. Write a definition for each boldface term in the paragraph above and for each of the following terms: pH, hypothesis, prediction, variable, control group. Use a separate sheet of paper. **Answers appear in the Teacher's Notes.**

2. Based on the objectives for this lab, write a question you would like to explore about the effect of acid rain, for example, When is a plant most susceptible to acid rain?

   Answers will vary.
   _____

   _____

Procedure
PART A: DESIGN AN EXPERIMENT

1. Work with members of your lab group to explore one of the questions written for step 2 of Before You Begin. To explore the question, design an experiment that uses the materials listed for this lab.

   > **You Choose**
   > As you design your experiment, decide the following:
   >   **a.** what question you will explore
   >   **b.** what hypothesis you will test
   >   **c.** how to simulate growing seeds in soil moistened by acid rain
   >   **d.** how to keep seeds moist during the experiment
   >   **e.** what your test solutions and control will be
   >   **f.** how to measure seedling growth
   >   **g.** what to record in your data table

2. Write a procedure for your experiment. Make a list of all the safety precautions you will take. Have your teacher approve your procedure and safety precautions before you begin the experiment.

Name _____ Class _____ Date _____

Observing the Effects of Acid Rain on Seeds *continued*

PART B: CONDUCT YOUR EXPERIMENT

3. Put on safety goggles, protective gloves, and a lab apron.

4. Place your seeds in a 250 mL beaker, and slowly add enough mold inhibitor to cover the seeds. **CAUTION: The mold inhibitor contains household bleach, which is a base.** Soak the seeds for 10 minutes, and then pour the mold inhibitor into the proper waste container. Gently rinse the seeds with distilled water, and place them on clean paper towels.

5. Set up your group's experiment. **CAUTION: Solutions with a pH below 7.0 are acids.** Conduct your experiment for 7–10 days. Make observations every 1–2 days, and note any changes. Record each day's observations in the data table below.

DATA TABLE		
Solution	**Date**	**Observations**

PART C: CLEANUP AND DISPOSAL

6. Dispose of solutions, broken glass, and seeds in the designated waste containers. Do not pour chemicals down the drain or put lab materials in the trash unless your teacher tells you to do so.

7. Clean up your work area and all lab equipment. Return lab equipment to its proper place. Wash your hands thoroughly before you leave the lab and after you finish all work.

ANALYZE AND CONCLUDE

1. Summarizing Results Describe any changes in the look of your seeds during the experiment. Discuss seed type, average seed size, number of germinated seeds, and changes in seedling length.

Answers will vary. _____

_____

_____

Name _____ Class _____ Date _____

Observing the Effects of Acid Rain on Seeds *continued*

2. Analyzing Results Were there any differences between the solutions? Explain.

Answers will vary. Plants usually grow best in a pH of 4.5 to 6.5.

_____

3. Analyzing Methods What was the control group in your experiment?

The control group was the seeds germinated in distilled water.

4. Analyzing Data Use graph paper to make graphs of your group's data. Plot seedling growth (in millimeters) on the y-axis. Plot number of days on the x-axis.

5. Relating Concepts What scientific methods did you use to design and conduct your experiment?

Answers should include collecting observations, asking questions, forming

hypotheses and making predictions, confirming predictions with experi-

ments, and drawing conclusions.

6. Evaluating Methods How could your experiment be improved?

Answers will vary. For example, students may suggest increasing the sample

size in each group.

7. Inferring Conclusions How do acidic conditions appear to affect seeds?

Answers will vary. In general, very acidic conditions inhibit seedling growth.

_____

8. Predicting Outcomes How might acid rain affect the plants in an ecosystem?

Answers will vary. Acidic rain might inhibit plant growth or kill plants.

_____

9. Further Inquiry Write a new question about the effect of acid rain that could be explored with another investigation.

Answers will vary. For example: What are the effects of acidic solutions on

mature plants?

Observing Enzyme Detergents

Teacher Notes

TIME REQUIRED 45 minutes for Day 1; 30 minutes for Day 2; and 15 minutes for Day 3.

SKILLS ACQUIRED

Designing Experiments

Measuring

Organizing and Analyzing Data

RATING

Easy ←— 1   2   3   4 —→ Hard

Teacher Prep–1

Student Setup–2

Concept Level–2

Cleanup–1

SCIENTIFIC METHODS

In this lab, students will:

Make Observations

Analyze the Results

Draw Conclusions

MATERIALS

Materials for this lab can be ordered from WARD'S. Use the Lab Materials QuickList Software on the **One-Stop Planner CD-ROM** for catalog numbers and to create a customized list of materials for this lab.

SAFETY CAUTION

Make sure students wear safety goggles and a lab apron. Caution students to avoid burns by working carefully when heating and pouring boiling water.

TIPS AND TRICKS

Before starting the lab, ask students why enzymes are added to detergents. (to help break down proteins and other substances from foods that may stain clothing) Ask students why detergent enzymes are stable during the hot water cycle. (The enzymes found in commercial laundry soap have been genetically engineered.) The rate of gelatin hydrolysis is slower in instant gelatin that contains sugar than in sugar-free gelatin. Students should bring in labeled samples (3–5 tablespoons) of each of five different laundry detergents. Labels should include active ingredients listed on the container. Students should have a control test tube with 15 drops (1 mL) of water (no detergent) added to the gelatin surface. Require -students to present a written procedure for their experiment and a list of all safety precautions before allowing them to gather materials for the lab. When students add the detergent solutions to the gelatin, the mixture will foam. During this reaction, carbon dioxide gas is released. The addition of washing soda (Na_2CO_3) to the gelatin raises the pH of the gelatin from 4 to 8, which is the optimum pH for protease activation in the detergent samples. Have students use a wax pencil to mark the test tubes at the uppermost level of the cooled gelatin in each tube. They will use this mark to measure the hydrolysis of the gelatin each day. Label the test tubes 1–6. To prepare a 10 percent solution of laundry detergent, students should dissolve 1 g of detergent in 9 mL of distilled water. Have students record the pH for each numbered detergent sample. Students can measure protein hydrolysis after 24 hours by using a wax pencil to draw a second line at the top of the gelatin layer and then measuring the distance (in mm) between the first line and the second line. This indicates the amount of hydrolysis of the protein in the gelatin by the enzymes in the detergent.

ANSWERS TO BEFORE YOU BEGIN

1. *enzymes*—molecules (usually proteins) that speed up chemical reactions

   pH—indicates the hydrogen ion concentration of a solution

   substrate—a reactant in a reaction that binds to an enzyme's active site

   protease—an enzyme that aids in the breakdown of proteins

2. Answers will vary. For example: How can you tell if a detergent contains enzymes?

Exploration Lab

Observing Enzyme Detergents

SKILLS

- Using scientific methods
- Measuring volume, mass, and pH

OBJECTIVES

- **Recognize** the function of enzymes in laundry detergents.
- **Relate** temperature and pH to the activity of enzymes.

MATERIALS

- safety goggles and lab apron
- balance
- graduated cylinder
- glass stirring rod
- 150 mL beaker
- 18 g regular instant gelatin or 1.8 g sugar-free instant gelatin
- 0.7 g Na_2CO_3
- tongs or a hot mitt
- 50 mL boiling water
- thermometer
- pH paper
- 6 test tubes
- test-tube rack
- pipet with bulb
- plastic wrap
- tape
- 50 mL beakers (6)
- 50 mL distilled water
- 1 g each of 5 brands of laundry detergent
- wax pencil
- metric ruler

SAFETY

 CAUTION: Always wear safety goggles and a lab apron to protect your eyes and clothing.

 CAUTION: Do not touch or taste any chemicals. Know the location of the emergency shower and eyewash station and how to use them. If you get a chemical on your skin or clothing, wash it off at the sink while calling to the teacher. Notify the teacher of a spill. Spills should be cleaned up promptly, according to your teacher's directions.

CAUTION: Glassware is fragile. Notify the teacher of broken glass or cuts. Do not clean up broken glass or spills with broken glass unless the teacher tells you to do so.

Before You Begin

Enzymes are substances that speed up chemical reactions. Each enzyme operates best at a particular **pH** and temperature. Substances on which enzymes act are called **substrates.** Many enzymes are named for their substrates. For example, a **protease** is an enzyme that helps break down proteins. In this lab, you will investigate the effectiveness of laundry detergents that contain enzymes.

1. Write a definition for each boldface term in the paragraph above. Use a separate sheet of paper. **Answers appear in the Teacher's Notes.**

2. Based on the objectives for this lab, write a question you would like to explore about enzyme detergents.

    Answers will vary. For example: How can you tell if a detergent contains

    enzymes?

Procedure

PART A: MAKE A PROTEIN SUBSTRATE

1. Put on safety goggles and a lab apron.

2. **CAUTION: Use tongs or a hot mitt to handle heated glassware.** Put 18 g of regular (1.8 g of sugar-free) instant gelatin in a 150 mL beaker. Slowly add 50 mL of boiling water to the beaker, and stir the mixture with a stirring rod. Test and record the pH of this solution.

3. Very slowly add 0.7 g of Na_2CO_3 to the hot gelatin while stirring. Note any reaction. Test and record the pH of this solution.

4. Place 6 test tubes in a test-tube rack. Pour 5 mL of the gelatin-Na_2CO_3 mixture into each tube. Use a pipet to remove any bubbles from the surface of the mixture in each tube. Cover the tubes tightly with plastic wrap and tape.

Observing Enzyme Detergents *continued*

Cool the tubes, and store them at room temperature until you begin Part C. Complete step 12.

PART B: DESIGN AN EXPERIMENT

5. Work with members of your lab group to explore one of the questions written for step 2 of **Before You Begin.** To explore the question, design an experiment that uses the materials listed for this lab.

> **You Choose**
>
> As you design your experiment, decide the following:
>
> **a.** what question you will explore
>
> **b.** what hypothesis you will test
>
> **c.** what detergent samples you will test
>
> **d.** what your control will be
>
> **e.** how much of each solution to use for each test
>
> **f.** how to determine if protein is breaking down
>
> **g.** what data to record in your data table

6. Write a procedure for your experiment. Make a list of all the safety precautions you will take. Have your teacher approve your procedure and safety precautions before you begin the experiment.

PART C: CONDUCT YOUR EXPERIMENT

7. Put on safety goggles and a lab apron.

8. Make a 10 percent solution of each laundry detergent by dissolving 1 g of detergent in 9 mL of distilled water.

9. Set up your experiment. Repeat step 12.

10. Record your data after 24 hours.

PART D: CLEANUP AND DISPOSAL

11. Dispose of solutions, broken glass, and gelatin in the designated waste containers. Do not pour chemicals down the drain or put lab materials in the trash unless your teacher tells you to do so.

12. Clean up your work area and all lab equipment. Return lab equipment to its proper place. Wash your hands thoroughly before leaving the lab and after finishing all work.

Observing Enzyme Detergents *continued*

Analyze and Conclude

1. **Analyzing Methods** Suggest a reason for adding Na_2CO_3 to the gelatin solution.

   The washing soda increases the pH of the gelatin from 4 to 8—the optimum

   pH for enzyme activity in this reaction.

2. **Analyzing Results** Make a bar graph of your data. Plot the amount of gelatin broken down (change in the depth of the gelatin) on the *y*-axis and detergent on the *x*-axis. Use a separate sheet of graph paper.

3. **Inferring Conclusions** What conclusions did your group infer from the results? Explain.

   Answers will vary. For example, enzymes operate best at a certain

   temperature and pH.

4. **Further Inquiry** Write a new question about enzyme detergents that could be explored with another investigation.

   Answers will vary. For example: Are enzymes in detergent stable in the

   presence of bleach?

Studying Animal Cells and Plant Cells

Teacher Notes

TIME REQUIRED One 45-minute class period

SKILLS ACQUIRED
Classifying
Communicating
Identifying/Recognizing Patterns
Inferring
Interpreting

RATING
Easy $\longleftarrow$ 1   2   3   4 $\longrightarrow$ Hard

Teacher Prep–3
Student Setup–2
Concept Level–2
Cleanup–3

SCIENTIFIC METHODS

In this lab, students will:
Make Observations
Draw Conclusions
Communicate the Results

MATERIALS

Materials for this lab can be ordered from WARD'S. Use the Lab Materials QuickList Software on the **One-Stop Planner CD-ROM** for catalog numbers and to create a customized list of materials for this lab. Combine all wastes containing Lugol's solution. To this mixture add a few drops of a strong acid, such as 1.0 m sulfuric acid (H_2SO_4), to make the mixture slightly acidic. Then slowly add 0.1 m sodium thiosulfate ($Na_2S_2O_3$) while stirring until the mixture loses its yellowish-orange color. Pour the resulting mixture down the drain.

SAFETY CAUTIONS

Review all safety symbols and caution statements with students before beginning the lab. Make sure students wear safety goggles and a lab apron. Caution students that they must wear safety goggles when working with Lugol's iodine solution because iodine can cause severe eye damage. If students are using microscopes with mirrors, caution them not to look directly at the reflected light from the mirror.

TIPS AND TRICKS

Set up labeled containers for the disposal of broken glass and materials stained with Lugol's solution. Place enough materials for one class on a supply table. Allow students to take off their safety goggles when looking through the microscope. Have students use a scale of 1 μm = 51 cm when drawing cells. Help students take measurements of cell size.

ANSWERS TO BEFORE YOU BEGIN

1. *light microscope*—a microscope that uses a beam of light passing through one or more lenses; *cytoplassm*—the material inside a cell between the cell membrane and the nuclearmembrane; *cell membrane*—a lipid bilayer with embedded proteins that encloses a cell; *nucleus*—an organelle found only in eukaryotic cells that houses the DNA; *nucleolus*—a structure in the nucleus of eukaryotic cells in which ribosomal proteins form; *vacuole*—a membrane-enclosed sac that contains substances, such as nutrients, needed by the cell; *cell wall*—a protective layer outside the cell membrane found in cells of bacteria, plants, fungi, and some protists; *chloroplast*—an organelle found only in the cells of plants and photosynthetic protists that uses light energy to drive the synthesis of organic compounds from carbon dioxide and water; *endoplasmic reticulum*—an extensive system of internal membranes that move proteins and other substances through the cell

2. Because many cell structures are transparent, a stain may be needed to distinguish certain cell parts. Additionally, some stains coat cell parts, such as flagella, making them thicker and easier to see.

3. Answers will vary. For example, how does the size of animal cells compare with the size of plant cells?

Skills Practice Lab

Studying Animal Cells and Plant Cells

SKILLS

- Using a compound microscope
- Drawing

OBJECTIVES

- **Identify** the structures you can see in animal cells and plant cells.
- **Compare** and **contrast** the structure of animal cells and plant cells.

MATERIALS

- compound light microscope
- prepared slide of human epithelial cells
- safety goggles
- lab apron
- polyethylene gloves
- sprig of *Elodea*
- forceps
- microscope slides and cover slips
- dropper bottle of Lugol's iodine solution

SAFETY

  CAUTION: Always wear safety goggles and a lab apron to protect your eyes and clothing.

CAUTION: Do not touch or taste any chemicals. Know the location of the emergency shower and eyewash station and how to use them. If you get a chemical on your skin or clothing, wash it off at the sink while calling the teacher. Notify the teacher of a spill. Spills should be cleaned up promptly, according to your teacher's directions.

 CAUTION: Glassware is fragile. Notify the teacher of broken glass or cuts. Do not clean up broken glass or spills with broken glass unless the teacher tells you to do so.

Before You Begin

You can see many cell parts with a **light microscope.** In animal cells, the **cytoplasm, cell membrane, nucleus, nucleolus,** and **vacuoles** can be seen. In plants cells, the **cell wall** and **chloroplasts** can also be be seen. Stains add color to cell parts and make them more visible with a light microscope. A stain can even make the **endoplasmic reticulum** visible. In this lab, you will use a light microscope to examine animal and plant cells.

1. Write a definition for each boldface term in the paragraph above. Use a separate sheet of paper. **Answers appear in the Teacher's Notes.**

2. Why might a stain be needed to see cell parts under a microscope?

Because many cell structures are transparent, a stain may be needed to

distinguish certain cell parts. Additionally, some stains coat cell parts, such

as flagella, making them thicker and easier to see.

3. Based on the objectives for this lab, write a question you would like to explore about cell structure.

Answers will vary. For example, how does the size of animal cells compare

with the size of plant cells?

Procedure

PART A: ANIMAL CELLS

1. 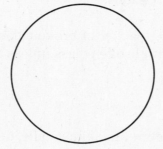 Examine a prepared slide of human epithelial cells under low power with a compound light microscope. Find cells that are separate from each other, and place them in the center of the field of view. Switch to high power, and adjust the diaphragm until you can see the cells more clearly. Identify as many cell parts as you can. *Note: Remember to use only the fine adjustment to focus at high power.*

2. Draw two or three epithelial cells as they look under high power. Label the cell membrane, the cytoplasm, the nuclear envelope, and the nucleus of at least one of the cells. Make a second drawing of these cells as you imagine they might look in the lining of your mouth.

Epithelial cells under high power

Epithelial cells in lining of mouth

Studying Animal Cells and Plant Cells *continued*

PART B: PLANT CELLS

3. 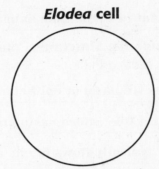 Using forceps, carefully remove a small leaf from near the top of an *Elodea* sprig. Place the whole leaf in a drop of water on a slide, and add a cover slip.

4. Observe the leaf under low power. Look for an area of the leaf in which you can see the cells clearly, and move the slide so that this area is in the center of the field of view. Switch to high power, and, if necessary, adjust the diaphragm. Identify as many cell parts as you can.

5. Find an *Elodea* cell in which you can see the chloroplasts clearly. Draw this cell. Label the cell wall, a chloroplast, and any other cell parts that you can see.

Elodea cell

6. Notice if the chloroplasts are moving in any of the cells. If you do not see movement, warm the slide in your hand or under a bright lamp for a minute or two. Look for movement of the cell contents again under the high power. Such movement is called cytoplasmic streaming.

7.  Put on safety goggles and a lab apron. Make a wet mount of another *Elodea* leaf, using Lugol's iodine solution instead of water. **CAUTION: Lugol's solution stains skin and clothing. Promptly wash off spills.** Observe these cells under low and high power.

8. Draw a stained *Elodea* cell. Label the cell wall and a chloroplast, as well as the central vacuole, the nucleus, and the cell membrane if they are visible.

Stained *Elodea* cell

PART C: CLEANUP AND DISPOSAL

9. Dispose of solutions, broken glass, and *Elodea* leaves in the waste containers designated by your teacher. Do not pour chemicals down the drain or put lab materials in the trash unless your teacher tells you to do so.

10. Clean up your work area and all lab equipment. Return lab equipment to its proper place. Wash your hands thoroughly before you leave the lab and after you finish all work.

Analyze and Conclude

1. Recognizing Patterns In what observable ways are animal and plant cells similar in structure, and in what observable ways are they different?

Animal and plant cells both have cytoplasm and a nucleus. Both types of

cells are bounded by a cell membrane. Unlike animal cells, plant cells have

chloroplasts, a cell wall, and central vacuole. Plant cells are also more con-

sistent in shape than animal cells.

2. Comparing Structures Compare and contrast the cytoplasm of epithelial cells and *Elodea* cells.

The cytoplasm of epithelial cells appears grainy. The cytoplasm of Elodea has

observable chloroplasts present, is usually located near the edge of the cell,

and typically flows in one direction.

3. Analyzing Methods What is the reason for staining *Elodea* cells with iodine?

The stain makes the nucleus, cell wall, and nucleolus more visible.

4. Inferring Conclusions Lugol's iodine solution causes the movement of chloro-plasts to stop. Explain why.

Lugol's solution is a poison that kills the cell and therefore stops the chloro-

plasts from moving.

5. Inferring Conclusions If some of the epithelial cells were folded over on themselves but were still transparent, what could you conclude about their thickness?

A reasonable conclusion is that epithelial cells are extremely thin.

6. Further Inquiry Write a new question about cell structure that could be explored with another investigation.

Answers will vary. For example: How do other animal and plant cells

compare?

Analyzing the Effect of Cell Size on Diffusion

Teacher Notes

TIME REQUIRED One 45-minute period

SKILLS ACQUIRED
Constructing Models
Designing Experiments
Inferring
Organizing and Analyzing Data

RATING
Easy ← 1   2   3   4 → Hard

Teacher Prep–2
Student Setup–3
Concept Level–1
Cleanup–1

SCIENTIFIC METHODS

In this lab, students will:
Make Observations
Analyze the Results
Draw Conclusions

MATERIALS

Materials for this lab can be ordered from WARD'S. Use the Lab Materials QuickList Software on the **One-Stop Planner CD-ROM** for catalog numbers and to create a customized list of materials for this lab.

SAFETY CAUTIONS

Make sure that students wear safety goggles and a lab apron.

Phenolphthalein solutions are flammable, and the vapors can explode when mixed with air. Make sure that there are no flames or sources of ignition, such as sparks, when you are using the phenolphthalein solution.

Remind students to read the Safety section before beginning the lab.

TIPS AND TRICKS

Phenolphthalein agar can be prepared in the laboratory. Add drops of 0.1 M sodium hydroxide solution (prepared by using 4 g NaOH diluted to 1L) to turn the agar red. Pour the mixture into a flat pan to a depth of slightly more than 3 cm. After the agar hardens, cut it into $3 \times 3 \times 6$ cm rectangles. Place enough materials for a class on a supply table. Set up labeled containers for the disposal of solutions, broken glass, and phenolphthalein agar. Before allowing students to gather materials for the lab, have them present a written procedure for their experiment as well as a list of all safety precautions. Make sure that students rinse off their knives between each cutting of different cubes to prevent vinegar from the previous cube from contaminating the next cube. For disposal, wrap the phenolphthalein agar in newspaper, then place the newspaper in the garbage. Dilute the vinegar with water, then pour the vinegar down the sink.

ANSWERS TO BEFORE YOU BEGIN

1. *diffusion*—the random movement of particles of a substance from an area of high concentration to an area of lower concentration; *surface area-to-volume ratio*—the ratio of the surface area of an object to its volume; *surface area*—the area of the exterior surface of an object; *volume*—the amount of space an object takes up

2. Answers will vary. For example: How does surface area-to-volume ratio affect the diffusion of a substance into a cell?

Exploration Lab

Analyzing the Effect of Cell Size on Diffusion

SKILLS

- Using scientific methods
- Collecting, organizing, and graphing data

OBJECTIVES

- **Relate** the size of a cell to its surface area-to-volume ratio.
- **Predict** how the surface area-to-volume ratio of a cell will affect the diffusion of substances into the cell.

MATERIALS

- safety goggles
- lab apron
- disposable gloves
- block of phenolphthalein agar (3 × 3 × 6 cm)
- plastic knife
- metric ruler
- 250 mL beaker
- 150 mL of vinegar
- plastic spoon
- paper towel

SAFETY

  CAUTION: Always wear safety goggles and a lab apron to protect your eyes and clothing.

 CAUTION: Do not touch or taste any chemicals. Know the location of the emergency shower and eyewash station and how to use them. If you get a chemical on your skin or clothing, wash it off at the sink while calling to the teacher. Notify the teacher of a spill. Spills should be cleaned up promptly, according to your teacher's directions.

CAUTION: Glassware is fragile. Notify the teacher of broken glass or cuts. Do not clean up broken glass or spills with broken glass unless the teacher tells you to do so.

Analyzing the Effect of Cell Size on Diffusion *continued*

Before You Begin

Substances enter and leave a cell in several ways, including by **diffusion.** How efficiently a cell can exchange substances depends on the **surface area-to-volume ratio** (surface area ÷ volume) of the cell. **Surface area** is the size of the outside of an object. **Volume** is the amount of space an object takes up. In this lab, you will investigate how cell size affects the diffusion of substances into a cell. To do this, you will make cell models using agar that contains an indicator. This indicator will change color when an acidic solution diffuses into it.

1. Write a definition for each boldface term in the paragraph above. Use a separate sheet of paper. **Answers appear in the Teacher's Notes.**

2. Based on the objectives for this lab, write a question you would like to explore about cell size and diffusion.

   Answers will vary. For example: How does surface area-to-volume ratio

   affect the diffusion of a substance into a cell?

Procedure

PART A: DESIGN AN EXPERIMENT

1. Work with members of your lab group to explore one of the questions written for step 2 of **Before You Begin.** To explore the question, design an experiment that uses the materials listed for this lab.

   > **You Choose**
   >
   > As you design your experiment, decide the following:
   > **a.** what question you will explore
   > **b.** what hypothesis you will test
   > **c.** how many "cells" (agar cubes) you will have and what sizes they will be
   > **d.** how long to leave the "cells" in the vinegar
   > **e.** how to determine how far the vinegar diffused into a "cell"
   > **f.** how to prevent contamination of agar cubes as you handle them
   > **g.** what data to record in your data table

2. Write a procedure for your experiment. Make a list of all the safety precautions you will take. Have your teacher approve your procedure and safety precautions before you begin the experiment.

PART B: CONDUCT YOUR EXPERIMENT

3. Put on safety goggles, a lab apron, and disposable gloves.

4. ◆◆◆ Carry out the experiment you designed. Record your observations in your data table.

PART C: CLEANUP AND DISPOSAL

5. Dispose of solutions, broken glass, and agar in the designated waste containers. Do not pour chemicals down the drain or put lab materials in the trash unless your teacher tells you to do so.

6. Clean up your work area and all lab equipment. Return lab equipment to its proper place. Wash your hands thoroughly before you leave the lab and after you finish all work.

Analyze and Conclude

1. Summarizing Results Describe any changes in the appearance of the cubes.

Cubes should appear light pink near their surfaces.
_____

2. Summarizing Results Make a graph using your group's data. Plot "Diffusion Distance (mm)" on the vertical axis. Use graph paper to plot "Surface Area-to-Volume Ratio" on the horizontal axis.

The line plotted on the graph should be horizontal.
_____

3. Analyzing Results Using the graph you made in item 2, make a statement about the relationship between the surface area-to-volume ratio and the distance a substance diffuses.

The diffusion distance is the same regardless of the surface area-to-volume

ratio.

4. Summarizing Results Make a graph using your group's data. Use graph paper to plot "Rate of Diffusion (mm/min)" (distance vinegar moved ÷ time) on the vertical axis. Plot "Surface Area-to-Volume Ratio" on the horizontal axis.

The line plotted on the graph should be horizontal.
_____

5. Analyzing Results Using the graph you made in item 4, make a statement about the relationship between the surface area-to-volume ratio and the rate of diffusion of a substance.

The rate of diffusion is constant regardless of the surface area-to-volume

ratio.

Analyzing the Effect of Cell Size on Diffusion *continued*

6. Evaluating Methods In what ways do your agar models simplify or fail to simulate the features of real cells?

Agar models ignore the selective permeability of cell membranes, the role of

membrane proteins in facilitated diffusion and active transport, and other

mechanisms of cell transport.

7. Calculating Calculate the surface area and volume of a cube with a side length of 5 cm. Calculate the surface area and volume of a cube with a side length of 10 cm. Determine the surface area-to-volume ratio of each of these cubes. Which cube has the greater surface area-to-volume ratio?

For a cube with a side length of 5 cm: surface area = 150 cm^2, volume =

125 cm^3; surface area-to-volume ratio = 6:5. For a cube with side length of

10 cm: surface area = 600 cm^2; volume = 1,000 cm^3; surface area-to-volume

ratio = 3:5. The small cube has the greater surface area-to-volume ratio.

8. Evaluating Conclusions How does the size of a cell affect the diffusion of substances into the cell?

In a small cell, substances do not have to diffuse as far to reach the center

of the cell.

9. Further Inquiry Write a new question about cell size and diffusion that could be explored with another investigation.

Answers will vary. For example: How can the cell membrane be modified to

take in materials more efficiently?

Observing Oxygen Production from Photosynthesis

Teacher Notes

TIME REQUIRED One 50-minute class period and about 15 minutes per day for 5 days.

SKILLS ACQUIRED
Measuring
Collecting Data
Graphing

RATING

Easy ◄——— 1   2   3   4 ——► Hard

Teacher Prep–3
Student Setup–3
Concept Level–2
Cleanup–3

SCIENTIFIC METHODS

In this lab, students will:
Make Observations
Analyze the Results
Draw Conclusions
Communicate the Results

MATERIALS

Materials for this lab can be ordered from WARD'S. Use the Lab Materials QuickList Software on the **One-Stop Planner CD-ROM** for catalog numbers and to create a customized list of materials for this lab.

SAFETY CAUTIONS

Review all safety symbols with students before beginning this lab.

TIPS AND TRICKS

Elodea is a common aquarium plant and can be found at some pet stores and most stores that sell aquarium fish. A 5 percent solution of baking soda and water can be made by dissolving 5 g of baking soda in 95 mL of water. You may need to have students practice placing the test tube over the inverted funnel. It may take two or three tries to get the test tube over the funnel stem without letting any air into the tube. First, fill the test tube with the solution. Place your thumb over the opening tightly so no air can get in. Submerge your thumb and the mouth of the test tube underwater. Once the mouth of the test tube is underwater you can release your thumb and maneuver the test tube over the stem of the funnel. Be sure you have the *Elodea* in place under the funnel before you begin.

Skills Practice Lab

Observing Oxygen Production from Photosynthesis

SKILLS

- Measuring
- Collecting Data
- Graphing

OBJECTIVE

- **Measure** amount of oxygen produced by an *Elodea* sprig.

MATERIALS

- 500 mL of 5 percent baking-soda-and-water solution
- 600 mL beaker
- 20 cm long *Elodea* sprigs (2–3)
- glass funnel
- test tube
- metric ruler
- protective gloves

SAFETY

 CAUTION: Always wear safety goggles and a lab apron to protect your eyes and clothing.

 CAUTION: Glassware is fragile. Notify the teacher of broken glass or cuts. Do not clean up broken glass or spills with broken glass unless the teacher tells you to do so.

 CAUTION: Wear disposable polyethlene gloves when handling any plant. Do not eat any part of a plant or plant seed used in the lab. Wash hands thoroughly after handling any part of a plant.

Before You Begin

Plants use **photosynthesis** to produce food. One product of photosynthesis is oxygen. In this activity, you will observe the process of photosynthesis and determine the rate of photosynthesis for *Elodea*.

1. Write a definition for the boldface term above.

   Photosynthesis is the process by which plants use carbon dioxide and the

   energy from sunlight to make sugar and carbon dioxide.

2. You will be using the data table provided to record your data.

Observing Oxygen Production from Photosynthesis *continued*

Procedure

1. Add 450 mL of baking-soda-and-water solution to a beaker.

2. Put two or three sprigs of *Elodea* in the beaker. The baking soda will provide the *Elodea* with the carbon dioxide it needs for photosynthesis.

3. Place the wide end of the funnel over the *Elodea*. The end of the funnel with the small opening should be pointing up. The *Elodea* and the funnel should be completely under the solution.

4. Fill a test tube with the remaining baking-soda-and-water solution. Place your thumb over the end of the test tube. Turn the test tube upside-down, taking care that no air enters. Hold the opening of the test tube under the solution and place the test tube over the small end of the funnel. Try not to let any solution leak out of the test tube as you do this.

5. Place the beaker setup in a well-lit area near a lamp or in direct sunlight.

6. Record that there was 0 mm gas in the test tube on day 0. (If you were unable to place the test tube without getting air in the tube, measure the height of the column of air in the test tube in millimeters. Record this value for day 0.) In this lab, change in gas volume is indicated by a linear measurement expressed in millimeters.

Data Table		
Amount of Gas Present in the Test Tube		
Days of exposure to light	**Total amount of gas present (mm)**	**Amount of gas produced per day (mm)**
0		
1		
2		
3		
4		
5		

7. For days 1 through 5, measure the amount of gas in the test tube. Record the measurements in your data table under the heading, "Total amount of gas present (mm)."

8. Calculate the amount of gas produced each day by subtracting the amount of gas present on the previous day from the amount of gas present today. Record these amounts under the heading, "Amount of gas produced per day (mm)."

9. Plot the data from your table on a graph.

Analyze and Conclude

1. **Summarizing Results** Using information from your graph, describe what happened to the amount of gas in the test tube.

   Students' graphs should show a gradual increase in the amount of gas in the

   test tube.

2. **Analyzing Data** How much gas was produced in the test tube after day 5?

   Answers will vary according to variables in the classroom, such as the

   amount of light and temperature.

3. **Drawing Conclusions** Write the equation for photosynthesis. Explain each part of the equation. For example, what ingredients are necessary for photosynthesis to take place? What substances are produced by photosynthesis? What gas is produced that we need in order to live?

   $6CO_2 + 6H_2O +$ light energy $\rightarrow C_6H_{12}O_6 + 6O_2$. CO_2 is carbon dioxide and

   comes from the baking soda solution. H_2O is the water in the solution. Light

   energy comes from the sun. $C_6H_{12}O_6$ is sugar (glucose), and O_2 is oxygen.

   Photosynthesis produces sugar and oxygen. Plants produce oxygen as a

   byproduct of photosynthesis; oxygen is the gas that we breathe and cannot

   live without.

4. **Predicting Patterns** What may happen to the oxygen level if an animal, such as a snail, were put in the beaker with the *Elodea* sprig while the *Elodea* sprig was making oxygen?

   Sample answer: The snail would use some of the oxygen generated by the

   plant. This could decrease the oxygen level.

5. **Further Inquiry** Write a new question about photosynthesis that could be explored with another investigation.

   Answers will vary. Example: How would increasing the temperature affect

   the amount of oxygen produced by the plant?

Modeling Mitosis

Teacher Notes

TIME REQUIRED 45–55 minutes

SKILLS ACQUIRED
Constructing Models
Identifying/Recognizing Patterns
Inferring
Predicting

RATING

Teacher Prep–1
Student Setup–2
Concept Level–2
Cleanup–1

Easy ←——— 1   2   3   4 ——→ Hard

SCIENTIFIC METHODS

In this lab, students will:
Make Observations
Ask Questions
Draw Conclusions

MATERIALS

Materials for this lab can be ordered from WARD'S. Use the Lab Materials QuickList Software on the **One-Stop Planner CD-ROM** for catalog numbers and to create a customized list of materials for this lab.

SAFETY CAUTIONS

Discuss all safety symbols and caution statements with students.

TIPS AND TRICKS

For each model, students need four pipe cleaners of one color, four pipe cleaners of another color, four wooden beads, 90 cm of yarn, and 16 small white labels. You may want to cut the string before the start of the lab. Review mitosis before starting the lab. If students have difficulty in beginning their models, ask the following questions: Which of these materials would make the best cell membrane? (yarn) Which would make the best spindle fibers? (yarn) Which would make the best chromosomes? (pipe cleaners) Which would make the best centromeres? (beads) Ask students what the cells they make were like before the chromosomes were duplicated. Emphasize that each pair of chromatids represents one chromosome. Most of the supplies for this lab can be saved and used again in the next lab.

Modeling Mitosis *continued*

ANSWERS TO BEFORE YOU BEGIN

1. *cell cycle*—a repeating five-phase sequence of eukaryotic cellular growth and division; *mitosis*—the process during cell division in which the nucleus of a cell divides into two nuclei, each with the same number and kind of chromosomes; *nondisjunction*—the failure of two chromosomes or chromatids to separate during nuclear division; *mutation*—a change in the DNA of a gene or chromosome; *chromatid*—one of a pair of strands of DNA that make up a chromosome; *centromere*—a section of a chromosome where two chromatids are joined; *spindle fiber*—a structure made of microtubules that helps pull chromatids apart during cell division; *cytokinesis*—the process of dividing the cytoplasm of a cell

Name _____ Class _____ Date _____

Modeling Mitosis

SKILLS

- Modeling
- Using scientific methods

OBJECTIVES

- **Describe** the events that occur in each stage of mitosis.
- **Relate** mitosis to genetic continuity.

MATERIALS

- pipe cleaners of at least two different colors
- yarn
- wooden beads
- white labels
- scissors

Before You Begin

The cell cycle includes all of the phases in the life of a cell. The **cell cycle** is a repeating sequence of cellular growth and division during the life of an organism. Mitosis is one of the phases in the cell cycle. **Mitosis** is the process by which the material in a cell's nucleus is divided during cell reproduction. In this lab, you will build a model that will help you understand the events of mitosis. You can also use the model to demonstrate the effects of **nondisjunction** and **mutations.**

1. Write a definition for each boldface term in the paragraph above and for the following terms: chromatid, centromere, spindle fiber, cytokinesis. Use a separate sheet of paper. **Answers appear in the Teacher's Notes.**

2. Where in the human body do cells undergo mitosis?

   everywhere in the human body except in most nerve and skeletal muscle cells

3. How does a cell prepare to divide during interphase of the cell cycle?

   During interphase, the cell grows, duplicates its chromosomes and

   organelles, and assembles microtubules.

4. Based on the objectives for this lab, write a question you would like to explore about mitosis.

   Answers will vary. For example: How many chromosomes will each new

   nucleus have after mitosis has occurred?

Modeling Mitosis *continued*

Procedure

PART A: DESIGN A MODEL

1. Work with the members of your lab group to design a model of a cell that uses the materials listed for this lab. Be sure your model cell has at least two pairs of chromosomes and is about to undergo mitosis.

> **You Choose**
>
> As you design your model, decide the following:
>
> **a.** what question you will explore
>
> **b.** how to construct a cell membrane
>
> **c.** how to show that your cell is diploid
>
> **d.** how to show the locations of at least two genes on each chromosome
>
> **e.** how to show that chromosomes are duplicated before mitosis begins

2. Write out the plan for building your model. Have your teacher approve the plan before you begin building the model.

The cell model the students build may vary. For example: The students may

use the yarn to represent the cell membrane and spindle fibers, the pipe

cleaners to represent the chromosomes, the beads to represent the cen-

tromeres, and the labels to indicate the genes on the chromosomes.

3. Build the cell model your group designed. **CAUTION: Sharp or pointed objects can cause injury. Handle scissors carefully. Promptly notify your teacher of any injuries.** Use your model to demonstrate the phases of mitosis. Draw and label each phase you model. **Student models will vary.**

4. Use your model to explore one of the questions written for step 4 of **Before You Begin.** Describe the steps you took to explore the question.

Answers will vary.

_____

_____

_____

_____

_____

PART B: TEST HYPOTHESES

Answer each of the following questions by writing a hypothesis. Use your model to test each hypothesis, and describe your results.

5. Cytokinesis follows mitosis. How will the size of each new cell that is formed following cytokinesis compare with that of the original cell?

Each new cell will initially be smaller than the original cell that divides.

_____

6. Sometimes two chromatids fail to separate during mitosis. How might this failure affect the chromosome number of the two new cells?

Nondisjunction of one of the chromosomes will result in one of the new cells

having two copies of that chromosome and the other cell having none.

_____

7. A mutation is a permanent change in a gene or chromosome. What effect might a mutation in a parent cell have on future generations of cells that result from the parent cell?

All of the cells in the subsequent generations of cells will carry the

mutation.

PART C: CLEANUP AND DISPOSAL

8. Dispose of paper and yarn scraps in the designated waste container.

9. Clean up your work area and all lab equipment. Return lab equipment to its proper place. Wash your hands thoroughly before you leave the lab and after you finish all work.

Analyze and Conclude

1. Analyzing Results How do the nuclei you made by modeling mitosis compare with the nucleus of the model cell you started with? Explain your result.

The nuclei students made should be the same as the nucleus of the original

cell except that the chromosomes in the original cell are not replicated until

right before mitosis.

_____

Modeling Mitosis *continued*

2. Evaluating Methods How could you modify your model to better illustrate the process of mitosis?

Answers will vary. Students could mention that the pipe cleaners do not

show how the shapes of the chromosomes change as they are pulled apart.

_____

_____

3. Recognizing Patterns How does the genetic makeup of the cells that result from mitosis compare with the genetic makeup of the original cell?

The genes found in the cells that result from mitosis are the same as the

genes in the original cell. The chromosomes replicate before cell division.

One copy of each gene goes to each new cell.

_____

4. Inferring Conclusions How is mitosis important?

Answers will vary. Students should mention that mitosis preserves the

chromosome number and genetic makeup of cells.

_____

5. Further Inquiry Write a new question about mitosis or the cell cycle that could be explored with your model.

Answers will vary. For example: What happens if the DNA is not replicated

before mitosis begins?

_____

Modeling Meiosis

Teacher Notes

TIME REQUIRED One 45-minute class period

SKILLS ACQUIRED
Communicating
Constructing Models
Identifying/Recognizing Patterns
Interpreting
Predicting

RATING
Easy ◄——— 1   2   3   4 ———► Hard

Teacher Prep–1
Student Setup–1
Concept Level–2
Cleanup–1

SCIENTIFIC METHODS

In this lab, students will:

Make Observations   Analyze the Results
Ask Questions   Draw Conclusions
Test the Hypothesis   Communicate the Results

MATERIALS

Materials for this lab can be ordered from WARD'S. Use the Lab Materials QuickList Software on the **One-Stop Planner CD-ROM** for catalog numbers and to create a customized list of materials for this lab. For each model, students need 4 pipe cleaners of one color and 4 pipe cleaners of another color, 4 wooden beads, 90 cm of yarn, and 16 small white labels.

SAFETY CAUTIONS

Review all safety symbols with students before beginning the lab.

TIPS AND TRICKS

Review the stages of meiosis before beginning the lab. If students have difficulty in building their models, ask the following questions:

1. Which of these materials would make the best cell membrane? (yarn)

2. Which would make the best spindle fibers? (yarn)

3. Which would make the best chromosomes? (pipe cleaners)

4. Which would make the best centromeres? (beads)

Emphasize that each pair of chromatids represents one chromosome. Be sure students show that homologous chromosomes pair during metaphase I of meiosis.

Exploration Lab

Modeling Meiosis

SKILLS

- Modeling
- Using scientific methods

OBJECTIVES

- **Describe** the events that occur in each stage of the process of meiosis.
- **Relate** the process of meiosis to genetic variation.

MATERIALS

- pipe cleaners of at least two different colors
- yarn
- wooden beads
- white labels
- scissors

Before You Begin

Meiosis is the process that results in the production of cells with half the normal number of chromosomes. It occurs in all organisms that undergo **sexual reproduction.** In this lab, you will build a model that will help you understand the events of meiosis. You can also use the model to demonstrate the effects of events such as **crossing-over** to explain results such as **genetic recombination.**

1. Write a definition for following terms: homologous chromosomes, gamete.

   Homologous chromosomes are chromosomes that are the same size and

   shape and carry genes for the same traits. A gamete is a reproductive cell,

   either a sperm or an egg.

2. In what organs in the human body do cells undergo meiosis?

   Meiosis occurs in the ovaries and testes.

3. During interphase of the cell cycle, how does a cell prepare for dividing?

   A cell's chromosomes duplicate, and certain organelles replicate.

4. Based on the objectives for this lab, write a question you would like to explore about meiosis.

Answers will vary. For example: How many chromosomes will each new

nucleus have after meiosis has occurred?

Procedure
PART A: DESIGN A MODEL

1. Work with the members of your lab group to design a model of a cell using the materials listed for this lab. Be sure that your model cell has at least two pairs of chromosomes.

2. Use a separate sheet of paper to write out the plan for building your model. Have your teacher approve the plan before you begin building the model.

3. Build the cell model your group designed. **CAUTION: Sharp or pointed objects can cause injury. Handle scissors carefully.** Use your model to demonstrate the phases of meiosis. Draw and label each phase you model.

4. Use your model to explore one of the questions written by your group for step 4 of Before You Begin. Describe the steps you took to explore your question.

> **You Choose**
> As you design your experiment, decide the following:
> **a.** what question you will explore
> **b.** how to construct a cell membrane
> **c.** how to show that your cell is diploid
> **d.** how to show the locations of at least two genes on each chromosome
> **e.** how to show that chromosomes are duplicated before meiosis begins

PART B: TEST HYPOTHESES

Answer each of the following questions by writing a hypothesis. Use your model to test each hypothesis, and describe your results.

5. In humans, gametes (eggs and sperm) result from meiosis. Will all gametes produced by one parent be identical?

Hypotheses will vary. Students should find that the gametes will be identical

only if the parent is homozygous for every one of its genetic traits.

6. When an egg and a sperm fuse during sexual reproduction, the resulting cell (the first cell of a new organism) is called a zygote. How many copies of each chromosome and each gene will be found in a zygote?

Hypotheses will vary. Students should find two copies of each chromosome

and two copies of each gene in a zygote.

7. Crossing-over frequently occurs between the chromatids of homologous chromosomes during meiosis. Under what circumstances does crossing-over result in new combinations of genes in gametes?

Hypotheses will vary. Students should find that crossing-over can produce

new combinations of genes when an organism has different versions of the

genes on the parts that cross over.

8. Synapsis (the pairing of homologous chromosomes) must occur before crossing-over can take place. How would the outcome of meiosis be different if synapsis did not occur?

Hypotheses will vary. Students may find the wrong chromosome number in a

gamete if the homologous pairs do not separate properly during anaphase I.

Also, there would be less genetic variation.

PART C: CLEANUP AND DISPOSAL

9. Dispose of paper and yarn scraps in the designated waste container.

10. Clean up your work area and all lab equipment. Return lab equipment to its proper place. Wash your hands thoroughly before you leave the lab and after finishing all work.

Analyze and Conclude

1. Analyzing Results How do the nuclei you made by modeling meiosis compare with the nucleus of the cell you started with? Explain your result.

The nuclei made by meiosis have half the original chromosome number.

Two divisions of the nuclear material occur.

2. Recognizing Relationships How are homologous chromosomes different from chromatids?

Homologous chromosomes are the same size and have genes for the same

traits, but they are not identical, as are chromatids.

3. Forming Reasoned Opinions How is synapsis important to the outcome of meiosis? Explain.

Synapses ensure that each new cell will get one member of each pair of

homologous chromosomes.

4. Evaluating Methods How could you modify your model to better illustrate the process of meiosis?

Answers will vary. Students may suggest using different materials.

_____

_____

5. Drawing Conclusions How are the processes of meiosis similar to those of mitosis? How are they different?

Mitosis and meiosis are both forms of nuclear division. Both consist of

prophase, metaphase, anaphase, and telophase. Mitosis consists of one

division and results in two cells that are genetically the same and have the

same number of chromosomes as the original cell. Meiosis consists of two

divisions and results in four cells that each have half the number of chromo-

somes as the original cell and are not the same genetically.

6. Predicting Outcomes What would happen to the chromosome number of an organism's offspring if the gametes for sexual reproduction were made by mitosis instead of by meiosis?

The chromosome number would be twice that in its parents' cells.

_____

7. Further Inquiry Write a new question about meiosis or sexual reproduction that could be explored with your model.

Answers will vary. For example: How do mitosis and meiosis compare?

_____

_____

Modeling Monohybrid Crosses

Teacher Notes

TIME REQUIRED One 50-minute class period

SKILLS ACQUIRED
Predicting Outcomes
Calculating Data
Organizing Data
Analyzing Data

RATING
Easy $\xleftarrow{\hspace{1cm}}$ 1   2   3   4 $\xrightarrow{\hspace{1cm}}$ Hard

Teacher Prep–1
Student Setup–1
Concept Level–3
Cleanup–1

SCIENTIFIC METHODS

In this lab, students will:
Make Observations
Analyze the Results
Draw Conclusions
Communicate the Results

MATERIALS

Materials for this lab can be ordered from WARD'S. Use the Lab Materials QuickList Software on the **One-Stop Planner CD-ROM** for catalog numbers and to create a customized list of materials for this lab.

SAFETY CAUTIONS

Remind students to avoid tasting the peas and lentils.

ANSWERS TO BEFORE YOU BEGIN

1. A monohybrid cross is a genetic cross that involves one pair of contrasting traits. Alleles are different versions of a gene that code for the same trait. A dominant allele is a version of a gene that is expressed in the presence of other alleles that code for the same trait. A recessive allele is a version of a gene that is not expressed when it is in the presence of other alleles that code for the same trait.

2. Answers may vary.

PROCEDURE

8. Answers will vary but genotypic ratios should approximate 1 *GG*:2 *Gg*:1 *gg*. Phenotypic ratios should approximate 3:1, or three individuals with green seeds to one individual with yellow seeds.

9. See answer to Procedure step 8.

10. See answer to Procedure step 8.

11. See answer to Procedure step 8.

12. See answer to Procedure step 8.

Skills Practice Lab

Modeling Monohybrid Crosses

MATERIALS

- lentils
- green peas
- 2 Petri dishes

SKILLS

- Predicting outcomes
- Calculating data
- Organizing data
- Analyzing data

OBJECTIVES

- **Predict** the genotypic and phenotypic ratios of offspring resulting from the random pairing of gametes.
- **Calculate** the genotypic ratio and phenotypic ratio among the offspring of a monohybrid cross.

Before You Begin

A **monohybrid cross** is a cross that involves one pair of contrasting traits. Different versions of a gene are called **alleles.** When two different alleles are present and one is expressed completely and the other is not, the expressed allele is **dominant** and the unexpressed allele is **recessive.**

1. Write a definition for each boldface term in the paragraph above. Use a separate sheet of paper. **Answers appear in the Teacher's Notes.**

2. Based on the objectives for this lab, write a question you would like to explore about heredity.

   Answers may vary.

   _____

   _____

Procedure

PART A: SIMULATING A MONOHYBRID CROSS

1. You will model the random pairing of alleles by choosing lentils and peas from Petri dishes. These dried seeds will represent the alleles for seed color. A green pea will represent G, the dominant allele for green seeds, and a lentil will represent g, the recessive allele for yellow seeds.

Modeling Monohybrid Crosses *continued*

2. Each Petri dish will represent a parent. Label one Petri dish "female gametes" and the other Petri dish "male gametes." Place one green pea and one lentil in the Petri dish labeled "female gametes" and place one green pea and one lentil in the Petri dish labeled "male gametes."

3. Each parent contributes one allele to each offspring. Model a cross between these two parents by choosing a random pairing of the dried seeds from the two Petri dishes. Do this by simultaneously picking one seed from each Petri dish without looking. Place the pair of seeds together on the lab table. The pair of seeds represents the genotype of one offspring.

4. Record the genotype of the first offspring in Table A below.

Table A		
Gamete Pairings		
Trial	**Offspring genotype**	**Offspring phenotype**
1	**Answers will vary.**	**Answers will vary.**
2		
3		
4		
5		
6		
7		
8		
9		
10		

5. Return the seeds to their original dishes and repeat step 3 nine more times. Record the genotype of each offspring in Table A.

6. Based on each offspring's genotype, determine and record each offspring's phenotype.

GG = green seeds; *Gg* = green seeds; *gg* = yellow seeds.

_____

_____

PART B: CALCULATING GENOTYPIC AND PHENOTYPIC RATIOS

7. You will be using Table B to record your data.

8. Determine the genotypic and phenotypic ratios among the offspring. First count and record in Table B the number of homozygous dominant, heterozygous, and homozygous recessive individuals you recorded in Table A. Then

Modeling Monohybrid Crosses *continued*

record the number of offspring that produce green seeds and the number that produce yellow seeds under "Phenotypes" in Table B. **Answers to 8–12 appear in the Teacher's Notes.**

Table B		
Offspring Ratios		
Genotypes	**Total**	**Genotypic ratios**
Homozygous dominant *(GG)*		
Heterozygous *(Gg)*		_____ : _____ : _____
Homozygous recessive *(gg)*		
Phenotypes		**Phenotypic ratios**
Green seeds		
Yellow seeds		

9. Calculate the genotypic ratio for each genotype using the following equation:

$$\text{phenotypic ratio} = \frac{\text{number of offspring with a given genotype}}{\text{total number of offspring}}$$

10. Calculate the phenotypic ratio for each phenotype using the following equation:

$$\text{phenotypic ratio} = \frac{\text{number of offspring with a given phenotype}}{\text{total number of offspring}}$$

11. Now pool the data for the whole class, and record the data in Table C below.

12. Compare the class's sample with your small sample of 10. Calculate the genotypic and phenotypic ratios for the class data, and record them in Table C below.

13. Construct a Punnett square showing the parents and their offspring in your lab report.

14. Clean up your materials before leaving the lab.

Table C		
Offspring Ratios (Class Data)		
Genotypes	**Total**	**Genotypic ratios**
Homozygous dominant *(GG)*		
Heterozygous *(Gg)*		_____ : _____ : _____
Homozygous recessive *(gg)*		
Phenotypes		**Phenotypic ratios**
Green seeds		
Yellow seeds		

Modeling Monohybrid Crosses *continued*

Analyze and Conclude

1. **Summarizing Results** What character is being studied in this investigation?

The trait being investigated is seed, or fruit, color.

2. **Analyzing Data** What are the genotypes of the parents? Describe the genotypes of both parents using the terms *homozygous* or *heterozygous*, or both. Did Table B reflect a classic monohybrid-cross phenotypic ratio of 3:1?

Both parents are heterozygous green, *Gg*. Each seed represents an allele.

The pairs represent the gametes that an offspring will receive. Answers will

vary. In a sample size of 10 crosses, no clear ratio may be evident. When

combining data from the entire class, the 3:1 ratio should be seen.

3. **Drawing Conclusions** If a genotypic ratio of 1:2:1 is observed, what must the genotypes of both parents be?

Both parents must be heterozygous, *Gg*.

4. **Predicting Patterns** Show what the genotypes of the parents would be if 50 percent of the offspring were green and 50 percent of the offspring were yellow.

One parent would be heterozygous, *Gg*, and the other parent would be

homozygous recessive, *gg*.

5. **Further Inquiry** Construct a Punnett square for the cross of a heterozygous black guinea pig and an unknown guinea pig whose offspring include a recessive white-furred individual. Use a separate sheet of paper. What are the possible genotypes of the unknown parent?

Students should draw two Punnett squares. One should show a cross between two heterozygous individuals; the other should show a heterozygous individual crossed with a homozygous recessive individual. If *B* represents the dominant black fur color and *b* stands for white fur color, the unknown parent could be genotypically *Bb* or *bb*.

Modeling DNA Structure

Teacher Notes

TIME REQUIRED One 45-minute class period

SKILLS ACQUIRED
Communicating
Constructing Models
Identifying/Recognizing Patterns
Interpreting
Predicting

RATING
Easy ◄—— 1   2   3   4 ——► Hard

Teacher Prep–1
Student Setup–1
Concept Level–2
Cleanup–1

SCIENTIFIC METHODS

In this lab, students will:
Make Observations
Ask Questions
Test the Hypothesis
Analyze the Results
Draw Conclusions
Communicate the Results

MATERIALS

Materials for this lab can be ordered from WARD'S. Use the Lab Materials QuickList Software on the **One-Stop Planner CD-ROM** for catalog numbers and to create a customized list of materials for this lab. Obtain enough plastic soda straws to make 48 3-cm sections for each group. Students will also need 48 standard-size paper clips and 48 colored pushpins (start each group with 12 red, 12 blue, 12 yellow, and 12 green pushpins).

SAFETY CAUTIONS

Review all safety symbols with students before beginning the lab. Caution students to be careful of the sharp tips of the pushpins. Account for all materials at the end of the lab.

TIPS AND TRICKS

Let students know they may need more or less of each color as they decide on the sequence of their DNA. If desired, ask students to save the assembled "nucleotides" for use in other DNA modeling labs.

ANSWERS TO BEFORE YOU BEGIN

1. *DNA*—double-stranded, helical nucleic acid that stores hereditary information; *protein*—an organic molecule composed of linked amino acids; *nucleotides*—subunits of nucleic acids that consist of a nitrogenous base, a sugar, and a phosphate group; *double helix*—a spiral structure characteristic of the DNA molecule; *complementary*—characteristic of nucleic acids in which the sequence of bases on one strand determines the sequence of bases on the other; *replication*—the process by which DNA in a cell is copied; *mutation*—a change in the nucleotide sequence of a gene or chromosome.

Exploration Lab

Modeling DNA Structure

SKILLS

- Modeling
- Using scientific methods

OBJECTIVES

- **Design** and analyze a model of DNA.
- **Describe** how replication occurs.
- **Predict** the effect of errors during replication.

MATERIALS

- plastic soda straws, 3 cm sections
- metric ruler
- pushpins (red, blue, yellow, and green)
- paper clips

Before You Begin

DNA contains the instructions that cells need in order to make every **protein** required to carry out their activities and to survive. DNA is made of two strands of **nucleotides** twisted around each other in a **double helix.** The two strands are **complementary,** that is, the sequence of bases on one strand determines the sequence of bases on the other strand. The two strands are held together by hydrogen bonds.

In this lab, you will build a model to help you understand the structure of DNA. You can also use the DNA model to illustrate and explore processes such as **replication** and **mutation.**

1. Write a definition for each boldface term in the paragraphs above and for each of the following terms: replication fork, base-pairing rules. Use a separate sheet of paper. **Answers appear in the Teacher's Notes.**

2. Identify the three different components of a nucleotide.

   a nitrogen base, a sugar, a phosphate

   _____

3. Identify the four different nitrogen bases that can be found in DNA nucleotides.

   adenine, guanine, cytosine, and thymine

   _____

4. Based on the objectives for this lab, write a question you would like to explore about DNA structure.

Answers may vary. For example: Will the sequence of nucleotides on the two

strands of DNA be identical?

Procedure
PART A: DESIGN A MODEL

1. Work with the members of your lab group to design a model of DNA that uses the materials listed for this lab. Be sure that your model has at least 12 nucleotides on each strand.

You Choose

As you design your model, decide the following:

    a. what question you will explore

    b. how to use the straws, pushpins, and paper clips to represent the three components of a nucleotide

    c. how to link (bond) the nucleotides together

    d. in what order you will place the nucleotides on each strand

---

2. Write out the plan for building your model. Use a separate sheet of paper. Have your teacher approve the plan before you begin building the model.

3. Build the DNA model your group designed. **CAUTION: Sharp or pointed objects may cause injury. Handle pushpins carefully.** Sketch and label the parts of your DNA model.

4. Use your model to explore one of the questions written for step 4 of **Before You Begin.**

PART B: DNA REPLICATION

5. Discuss with your lab group how the model you built for Part A may be used to illustrate the process of replication.

6. Write a question you would like to explore about replication. Use your model to explore the question you wrote. On a separate sheet of paper, sketch and label the steps of replication.

Answers may vary. For example: What happens to the new DNA strand if the

wrong nucleotide is added during replication?

PART C: TEST HYPOTHESIS

Answer each of the following questions by writing a hypothesis. Use your model to test each hypothesis, and describe your results.

7. Mitosis follows replication. How might the cells produced by mitosis be affected if nucleotides on one DNA strand were incorrectly paired during replication?

Hypotheses will vary. The DNA in the new cells that result from cell division

will not be identical.

8. What would happen if only one strand in a DNA molecule were copied during replication?

Hypotheses will vary. The single strand would not have a strand to pair with

and would not be complete.

PART D: CLEANUP AND DISPOSAL

9. Dispose of damaged pushpins in the designated waste container.

10. Clean up your work area and all lab equipment. Return lab equipment to its proper place. Wash your hands thoroughly before you leave the lab and after you finish all work.

Analyze and Conclude

1. Analyzing Results In your original DNA model, were the two strands identical to each other?

No, they were complementary.

2. Relating Concepts How does DNA structure ensure that the two DNA molecules made by replication are the same as the original DNA molecule?

The DNA structure is such that the two strands are complementary to each

other.

3. Drawing Conclusions Did the two DNA molecules you made in step 6 have the same nitrogen-base sequence as your original model DNA molecule?

yes

4. Inferring Relationships The order of nitrogen bases on a DNA strand is a code for making proteins. What does this mean has happened to the "code" in one of the DNA molecules you made in step 7?

The code has been changed. _____

_____

5. Predicting Outcomes What would happen if the DNA in a cell that is about to divide were not replicated?

One of the resulting cells would die because it would not receive any pro-

tein-making instructions. _____

_____

6. Inferring Information What are the advantages of having DNA remain in the nucleus of a cell?

The nucleus provides an isolated and protected area for storing genetic

information. _____

_____

7. Further Inquiry Write a new question about DNA that could be explored with your model.

Answers may vary. For example: How do DNA molecules differ among various

species of animals and plants? _____

_____

Modeling Protein Synthesis

Teacher Notes

TIME REQUIRED One 45-minute class period

SKILLS ACQUIRED
Communicating
Constructing Models
Identifying/Recognizing Patterns
Interpreting

RATING
Easy ←——1———2———3———4——→ Hard

Teacher Prep–2
Student Setup–2
Concept Level–3
Cleanup–1

SCIENTIFIC METHODS

In this lab, students will:
Make Observations
Ask Questions
Draw Conclusions
Communicate the Results

MATERIALS

Materials for this lab can be ordered from WARD'S. Use the Lab Materials QuickList Software on the **One-Stop Planner CD-ROM** for catalog numbers and to create a customized list of materials for this lab.

SAFETY CAUTIONS

Discuss all safety symbols and caution statements with students. Caution students to be careful of the sharp tips of the pushpins.

TIPS AND TRICKS

Each group will need enough plastic soda straws to make about 40 sections of one color (3 cm each) and 15 sections of a different color. Each group will also need about 40 standard-sized paper clips, 40 colored pushpins, and enough materials to make at least eight tRNA/amino acid cards. Before beginning the lab, use Figures 2 and 5 of the chapter to review the steps in gene expression. Account for all materials at the end of the lab, and have students store the pushpins by color.

ANSWERS TO BEFORE YOU BEGIN

1. *protein*—an organic molecule made of amino acids; *protein synthesis*—the process by which proteins are made based on the information encoded in a gene; *mRNA*—an RNA copy of a gene that carries the instructions for making a protein from a gene and delivers it to a ribosome for translation; *amino acids*—organic molecules that are the building blocks of proteins; *mutation*—a change in the DNA of a gene or chromosome; *sickle cell anemia*—a condition caused by a mutant allele that produces a defective form of hemoglobin; *hemoglobin*—a protein found in red blood cells that binds with and carries oxygen throughout the body; *transcription*—a stage of gene expression in which the information in DNA is transferred to an RNA molecule; *translation*—a stage of gene expression in which the information in mRNA is used to make a protein; *tRNA*—an RNA molecule that temporarily carries a specific amino acid to a ribosome during translation; *ribosome*—a cytoplasmic organelle on which proteins are made; *codon*—a three-nucleotide sequence in mRNA that encodes an amino acid or signifies a start or stop signal; *anticodon*—a three-nucleotide sequence on tRNA that recognizes a complementary codon on mRNA.

Exploration Lab

Modeling Protein Synthesis

SKILLS

- Modeling
- Using scientific methods

OBJECTIVES

- **Compare** and **Contrast** the structure and function of DNA and RNA.
- **Model** protein synthesis.
- **Demonstrate** how a mutation can affect a protein.

MATERIALS

- masking tape
- plastic soda-straw pieces of one color
- plastic soda-straw pieces of a different color
- paper clips
- pushpins of five different colors
- marking pens of the same colors as the pushpins
- 3×5 in. note cards
- oval-shaped card
- transparent tape

Before You Begin

The nature of a **protein** is determined by the sequence of amino acids in its structure. During **protein synthesis**, the sequence of nitrogen bases in an **mRNA** molecule is used to assemble **amino acids** into a protein chain.

A mutation is a change in the nitrogen-base sequence of DNA. Many mutations lead to altered or defective proteins. For example, the genetic blood disorder **sickle cell anemia** is caused by a mutation in the gene for **hemoglobin**.

In this lab, you will build models that will help you understand how protein synthesis occurs. You can also use the models to explore how a mutation affects a protein.

1. Write a definition for each boldface term in the paragraph above and for each of the following terms: transcription, translation, tRNA, ribosome, codon, anticodon. Use a separate sheet of paper. **Answers appear in the Teacher's Notes.**

| Modeling Protein Synthesis *continued*

2. Describe three differences between DNA and RNA.

RNA is single stranded, and DNA is double stranded. RNA nucleotides con-

tain the sugar ribose and the nitrogen bases uracil, guanine, cytosine, or

adenine. DNA nucleotides contain the sugar deoxyribose and the nitrogen

bases thymine, guanine, cytosine, or adenine.

3. Based on the objectives for this lab, write a question you would like to explore about protein synthesis.

Answers will vary. For example: How does the cell's DNA determine the

proteins made by a cell?

Procedure

PART A: DESIGN A MODEL

1. Work with the members of your lab group to design models of DNA, RNA, and a cell. Use the materials listed for this lab.

> **You Choose**
>
> As you design your models, decide the following:
>
> **a.** what question you will explore
>
> **b.** how to represent DNA nucleotides
>
> **c.** how to represent RNA nucleotides
>
> **d.** how to represent five different nitrogen bases
>
> **e.** how to link (bond) nucleotides together
>
> **f.** how to represent tRNA molecules with amino acids
>
> **g.** how to represent the locations of DNA and ribosomes

2. Write out the plan for building your models. Have your teacher approve the plan before you begin building the models.

Answers may vary. Students might use straw segments to model the sugar-

phosphate backbone of DNA, different-colored pushpins to represent the

four different nitrogen bases, paper clips to represent the bonds that hold

the nucleotides in a chain, note cards to model tRNA molecules with amino

acids attached, and oval-shaped cards to represent ribosomes.

Modeling Protein Synthesis *continued*

3. Build the models your group designed. **CAUTION: Sharp or pointed objects may cause injury. Handle pushpins carefully.** Start your model of DNA with a strand of nucleotides that has the following sequence of nitrogen bases: TTTGGTCTCCTC.

PART B: MODEL PROTEIN SYNTHESIS

4. Use your models to demonstrate how transcription and translation occur. Draw and label the steps of each process on a separate sheet of paper. **Answers may vary. Check for students' understanding of each step in gene expression.**

5. Use your models to explore one of the questions written for step 3 of **Before You Begin.**

    Students' results will vary. Some students might produce a mutation and

    demonstrate how it affects protein synthesis. Some mutations cause the

    wrong amino acid to be inserted in the protein. Other mutations might insert

    a start or stop codon at an inappropriate place.

PART C: TEST HYPOTHESIS

Answer each of the following questions by writing a hypothesis. Use your models to test each hypothesis, and describe your results.

6. The DNA model you built for step 3 represents a portion of a gene for hemoglobin. Sickle cell anemia results from the substitution of an A for the T in the third codon of the nitrogen-base sequence given in step 3. How will this substitution affect a hemoglobin molecule?

    Hypotheses will vary. The substitution will change the amino acid sequence from

    lysine-proline-glutamic acid-glutamic acid to lysine-proline-valine-glutamic acid.

7. The addition of a nucleotide to a strand of DNA is a type of mutation called an *insertion*. What happens when an insertion occurs in the first codon in a DNA strand, before the DNA strand is transcribed?

    Hypotheses will vary. The codon triplets are shifted.

    _____

PART D: CLEANUP AND DISPOSAL

8. Dispose of damaged pushpins in the designated waste container.

9. Clean up your work area and all lab equipment. Return lab equipment to its proper place. Wash your hands thoroughly before you leave the lab and after you finish all work.

| Modeling Protein Synthesis *continued*

Analyze and Conclude

1. **Comparing Structures** How did the nitrogen-base sequence of the mRNA you made compare with that of the DNA it was transcribed from?

   They are complementary.
   _____

2. **Recognizing Relationships** How is the nitrogen-base sequence of a gene related to the structure of a protein?

   The nitrogen-base sequence of a gene contains a code that determines the

   amino-acid sequence of a protein.
   _____

3. **Recognizing Patterns** What is the relationship between the anticodon of a tRNA and the amino acid the tRNA carries?

   A tRNA with a particular anticodon always carries the same amino acid.
   _____

4. **Drawing Conclusions** How does a mutation in the gene for a protein affect the protein?

   A mutation can change the amino acid sequence of a protein, and thus may

   change its activity.
   _____

5. **Further Inquiry** Write a new question about protein synthesis that could be explored with your model.

   Example question: What happens if a mutation occurs that changes a codon

   to a stop codon?
   _____

Modeling Recombinant DNA

Teacher Notes

TIME REQUIRED 40–50 minutes

SKILLS ACQUIRED

Constructing Models
Identifying/Recognizing Patterns
Inferring
Interpreting
Predicting

RATING

Easy ← 1 2 3 4 → Hard

Teacher Prep–2
Student Setup–1
Concept Level–3
Cleanup–1

SCIENTIFIC METHODS

In this lab, students will:
Make Observations
Ask Questions
Analyze the Results
Draw Conclusions

MATERIALS

Materials for this lab can be ordered from WARD'S. Use the Lab Materials QuickList Software on the **One-Stop Planner CD-ROM** for catalog numbers and to create a customized list of materials for this lab.

SAFETY CAUTIONS

Review all safety symbols and caution statements with students before beginning the lab. Caution students to be careful of the sharp tips on the pushpins.

TIPS AND TRICKS

Make sure each group has enough materials before they begin the activity. Have students work in cooperative groups of four students each. Divide each group into pairs. One two-person team should complete steps 2–4 while the other team completes steps 5 and 6. The entire group should work together to complete steps 7–11. Before beginning the lab, review the structure of DNA and the steps of genetic engineering.

ANSWERS TO BEFORE YOU BEGIN

1. *genetic engineering*—the process of isolating a gene from the DNA of one organism and inserting it into the DNA of another organism

vector—agent, such as a virus or plasmid, used to carry a DNA fragment into a cell

plasmid—small circular DNA molecule, usually found in bacteria, that can replicate independently of the main chromosome

restriction enzyme—bacterial enzyme that cuts DNA at a specific sequence of nucleotides

sticky ends—single-stranded ends of DNA that are produced when a restriction enzyme cuts DNA

DNA ligase—enzyme that joins ends of DNA

recombinant DNA—DNA that contains DNA segments from different organisms

base-pairing rules—rules stating that in DNA, adenine on one strand always pairs with thymine on the opposite strand and cytosine on one strand always pairs with guanine on the opposite strand

| Exploration Lab |

Modeling Recombinant DNA

SKILLS

• Modeling

• Comparing

OBJECTIVES

• **Construct** a model that can be used to explore the process of genetic engineering.

• **Describe** how recombinant DNA is made.

MATERIALS

• paper clips (56)

• plastic soda straw pieces (56)

• pushpins (15 red, 15 green, 13 blue, and 13 yellow)

Before You Begin

Genetic engineering is the process of taking a gene from one organism and inserting it into the DNA of another organism. The gene is delivered by a **vector**, such as a virus, or a bacterial **plasmid.**

First, a fragment of a chromosome that contains the gene is isolated by using a **restriction enzyme**, which cuts DNA at a specific nucleotide-base sequence. Some restriction enzymes cut DNA unevenly, producing single-stranded **sticky ends.** The DNA of the vector is cut by the same restriction enzyme. Next, the chromosome fragment is mixed with the cut DNA of the vector. Finally, an enzyme called **DNA ligase** joins the ends of the two types of cut DNA, producing **recombinant DNA.**

In this lab, you will model genetic engineering techniques. You will simulate the making of recombinant DNA that has a human gene inserted into the DNA of a plasmid.

1. Write a definition for each boldface term in the paragraph above and for the term *base-pairing rules*. Use a separate sheet of paper.
   Answers appear in the Teacher's Notes.

2. Based on the objectives for this lab, write a question you would like to explore about the process of genetic engineering.

   Answers will vary. For example: What steps are involved in transferring a

   gene from one organism to another one?

Name _____ Class _____ Date _____

| Modeling Recombinant DNA *continued*

Procedure

PART A: MODEL GENETIC ENGINEERING

1. Make 56 model nucleotides. To make a nucleotide, insert a pushpin mid-way along the length of a 3 cm piece of a soda straw. **CAUTION: Handle pushpins carefully. Pointed objects can cause injury.** Push a paper clip into one end of the soda-straw piece until it touches the pushpin.

2. Begin a model of a bacterial plasmid by arranging nucleotides for one DNA strand in the following order: blue, red, green, yellow, red, red, blue, blue, green, red, blue, green, red, blue, blue, green, yellow, and red. Join two adjacent nucleotides by inserting the paper clip end of one into the open end of the other.

3. Using your first DNA strand and the base-pairing rules, build the complementary strand of plasmid DNA. **Note:** *Yellow is complementary to blue, and green is complementary to red.*

4. Complete your model of a circular plasmid by joining the opposite ends of each DNA strand. Make a sketch showing the sequence of bases in your model plasmid. Use the abbreviations B, Y, G, and R for the pushpin colors.

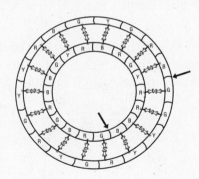

5. Begin a model of a human chromosome fragment made by a restriction enzyme. Place nucleotides for one DNA strand in the following order: BBRRYGGBRY. Build the second DNA strand by arranging the remaining nucleotides in the following order: BRRYGBYYGG.

6. Match the complementary portions of the two strands of DNA you made in step 5. Pair as many base pairs in a row as you can. Make a sketch showing the sequence of bases in your model of a human chromosome fragment.

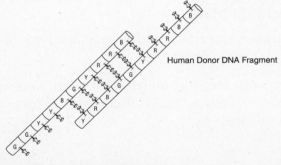

Human Donor DNA Fragment

Name _____ Class _____ Date _____

Modeling Recombinant DNA *continued*

7. Imagine that the restriction enzyme that cut the human chromosome fragment you made in steps 5 and 6 is moving around your model plasmid until it finds the sequence YRRBBG and its complementary sequence, BGGYYR. This restriction enzyme cuts each sequence between a B and a G. Find such a section in your sketch of your model plasmid's DNA.

The sketch should show the split at the nucleotides indicated by the arrows in item 4.

8. Simulate the action of the restriction enzyme on the section you identified in step 7. Open both strands of your model plasmid's DNA by pulling apart the adjacent green and blue nucleotides in each strand. Make a sketch of the split plasmid DNA molecule.

9. Move your model human DNA fragment into the break in your model plasmid's DNA molecule. Imagine that a ligase joins the ends of the human and plasmid DNA. Make a sketch of your final model DNA molecule.

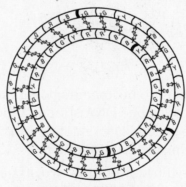

Final Bacterial DNA Molecule

PART B: CLEANUP AND DISPOSAL

10. Dispose of damaged pushpins in the designated waste container.

11. Clean up your work area and all lab equipment. Return lab equipment to its proper place. Wash your hands thoroughly before you leave the lab and after you finish all work.

Modeling Recombinant DNA *continued*

Analyze and Conclude

1. **Comparing Structures** Compare your models of plasmid DNA and human DNA.

   Both the plasmid DNA model and the human DNA model are double-

   stranded. The plasmid DNA model is circular and the human DNA model is

   linear. The human DNA model represents a fragment of a DNA molecule that

   has been cut with a restriction enzyme and has sticky ends.

2. **Relating Concepts** What do the sections of four unpaired nucleotides in your model human DNA fragment represent?

   sticky ends

3. **Comparing Structures** How did your original model plasmid DNA molecule differ from your final model DNA molecule?

   The original model plasmid DNA was smaller (had fewer nucleotides) and

   did not contain a gene from a human chromosome.

4. **Drawing Conclusions** What does the molecule you made in step 9 represent?

   Answers will vary. For example: What happens if the plasmid and human

   DNA do not have complementary sticky ends?

5. **Further Inquiry** Write a new question that could be explored with another investigation.

   recombinant DNA

Making a Timeline of Life on Earth

Teacher Notes

TIME REQUIRED 90 minutes

SKILLS ACQUIRED
Communicating
Identifying/Recognizing Patterns
Inferring
Measuring
Organizing and Analyzing Data

RATING
Easy $\longleftarrow$ 1   2   3   4 $\longrightarrow$ Hard

Teacher Prep–2
Student Setup–1
Concept Level–1
Cleanup–1

SCIENTIFIC METHODS

In this lab, students will:
Make Observations
Ask Questions
Draw Conclusions
Communicate the Results

MATERIALS

Materials for this lab can be ordered from WARD'S. Use the Lab Materials QuickList Software on the **One-Stop Planner CD-ROM** for catalog numbers and to create a customized list of materials for this lab.

TIPS AND TRICKS

Set up stations in various parts of the classroom with specimens, fossils, photographs, or slides of different types of organisms. Provide compound light microscopes if needed. During the first hour (or lab period), have students make their timelines and observe and record data for as many organisms as possible. During the second period, have students complete the investigation.

ANSWERS TO BEFORE YOU BEGIN

1. Definitions: *fossil*—preserved remains or traces of a past life form; *fossil record*—the record of past life-forms found in fossils; *timeline*—a series of events arranged in chronological order.

Exploration Lab

Making a Timeline of Life on Earth

SKILLS

- Observing
- Inferring relationships
- Organizing data

OBJECTIVES

- **Compare** and **contrast** the distinguishing characteristics of representative organisms of the six kingdoms.
- **Organize** the appearance of life on Earth in a timeline.

MATERIALS

- adding-machine tape (5 m roll)
- meterstick
- colored pens or pencils
- photographs or drawings of organisms from ancient Earth to present day

Before You Begin

About 4.5 billion years ago, Earth was a ball of molten rock. As the surface cooled, a rocky crust formed and water vapor in the atmosphere condensed to form rain. By 3.9 billion years ago, oceans covered much of Earth's surface. Rocks formed in these oceans contain **fossils** of bacterial cells that lived about 3.5 billion years ago. The **fossil record** shows a progression of life-forms and contains evidence of many changes in Earth's surface and atmosphere.

In this lab, you will make a **timeline** showing the major events in Earth's history and in the history of life on Earth, such as the evolution of new groups of organisms and the mass extinctions. This timeline can be used to study how living things have changed over time.

1. Write a definition for each boldface term in the paragraphs above. Use a separate sheet of paper. **Answers appear in the Teacher's Notes.**

2. Record your data in the data table provided.

3. Based on the objectives for this lab, write a question you would like to explore about the history of life on Earth.

   Answers will vary. For example: How did life on Earth change after oxygen

   entered the atmosphere?

Procedure

PART A: MAKING A TIMELINE

1. Make a mark every 20 cm along a 5 m length of adding-machine tape. Label one end of the tape "5 billion years ago" and the other end "Today." Write "20 cm = 200 million years" near the beginning of your timeline.

2. Locate and label a point representing the origin of Earth on your timeline. Use your textbook as a reference. Also locate and label the periods of the geologic time scale. **Check student timelines against the textbook.**

3. Using your textbook as a reference, mark the following events on your timeline: the first cyanobacteria appear; oxygen enters the atmosphere; the five mass extinctions; the first eukaryotes appear; the first multicellular organisms appear; the first vertebrates appear; the first plants, fungi, and land animals appear; the first dinosaurs and mammals appear; the first flowering plants appear; the first humans appear. **Check student answers against the textbook.**

4. Look at the photographs of organisms provided by your teacher. Identify the major characteristics of each organism. Record your observations in the data table below.

Sample Data:

Data Table		
Organism	**Kingdom**	**Characteristics/adaptation for life on Earth**
Insect (fossil in amber)	**Animal**	Segmented body, jointed legs, and wings
Fern (fossil mold in shale)	**Plant**	Large surface area to capture sunlight; reproductive structures on fronds

Making a Timeline of Life on Earth *continued*

5. Lay out your timeline on the floor in your classroom. Place photographs (or drawings) of the organisms you examined on your timeline to show when they appeared on Earth. **Check student answers against the textbook.**

6. Fold the timeline at the mark representing 4.8 billion years ago. This leaves 24 segments, each representing 200 million years on your timeline. Now you can think of each segment as 1 hour in a 24-hour day.

7. When you are finished, walk slowly along your timeline. Note the sequence of events in the history of life on Earth and the relative amount of time between each event.

PART B: CLEANUP AND DISPOSAL

8. Dispose of paper scraps in the designated waste container.

9. Clean up your work area and all lab equipment. Return lab equipment to its proper place.

ANALYZE AND CONCLUDE

1. Analyzing Information Think of each segment of your timeline as 1 hour in a 24-hour day as you answer each of the following questions.

a. How long has life existed on Earth?

Life on Earth has existed for at least 17.5 hours.

b. For what part of the day did only unicellular life-forms exist?

Only unicellular life-forms existed for about 14 hours.

c. At what time of day did the first plants appear on Earth?

The first plants appeared at about 9:30 P.M.

d. At what time of day did mammals appear on Earth?

Mammals appeared at about 11:00 P.M.

2. Summarizing Information Identify the major developments in life-forms that have occurred over the last 3.5 billion years.

Answers will vary. Life has changed from early prokaryotic, unicellular forms

to complex multicellular forms composed of eukaryotic cells. Many life-forms

that evolved have become extinct.

_____

3. Inferring Relationships How do mass extinctions appear to be related to the appearance of new major groups of organisms?

Mass extinctions appear to be followed closely by the evolution of many new types of organisms.

4. Justifying Conclusions Cyanobacteria are thought to be responsible for adding oxygen to Earth's atmosphere. Use your timeline to justify this conclusion.

Cyanobacteria appeared before the accumulation of oxygen in the atmosphere, and cyanobacteria produce oxygen. Therefore, it is reasonable to conclude that they are responsible for adding oxygen to Earth's atmosphere.

5. Calculating Determine the amount of time, as a percentage of the time that life has existed on Earth, that humans (Homo sapiens) have existed.

Humans have existed for 0.014 percent of the time that life has existed on Earth.

6. Further Inquiry Write a new question about the history of life on Earth that could be explored in another investigation.

Answers will vary. Sample answer: How have Earth's climates changed since Precambrian times?

Modeling Natural Selection

Teacher Notes

TIME REQUIRED Two 45-minute periods

SKILLS ACQUIRED
Collecting Data
Constructing Models
Experimenting
Organizing and Analyzing Data

RATING
Easy ←— 1 — 2 — 3 — 4 —→ Hard

Teacher Prep–2
Student Setup–2
Concept Level–3
Cleanup–1

SCIENTIFIC METHODS

In this lab, students will:
Make Observations
Ask Questions
Test the Hypothesis
Draw Conclusions

MATERIALS

Materials for this lab can be ordered from WARD'S. Use the Lab Materials QuickList Software on the **One-Stop Planner CD-ROM** for catalog numbers and to create a customized list of materials for this lab.

SAFETY CAUTIONS

Remind students to be careful with scissors and always cut in a direction away from the face and body.

TIPS AND TRICKS

Students will simulate the breeding of several generations of the birds and observe the effect of various phenotypes on the evolutionary success of these animals. The random nature of mutations is demonstrated by randomly changing the anterior and posterior wing position and wing circumference of the birds. Each student group will need one sheet of construction paper, a meter stick or tape measure, a pair of scissors, a straw, cellophane tape, a coin, and a die. To save time, have students create their table before beginning the lab, and have students work in pairs or small groups. You may wish to review and clarify the principles of selection. This investigation reinforces the concept of natural selection as differential reproduction rather than merely differential survival. Have students dispose of their paper scraps and paper birds in the trash.

ANSWERS TO BEFORE YOU BEGIN

1. *natural selection*—process by which populations change in response to their environment as individuals better adapted to the environment leave more offspring; *traits*—distinguishing characteristics; *genotype*—the genetic constitution of an organism as indicated by its set of alleles

Exploration Lab

Modeling Natural Selection

SKILLS
- Modeling a process
- Inferring relationships

OBJECTIVES
- **Model** the process of selection.
- **Relate** favorable mutations to selection and evolution.

MATERIALS
- scissors
- construction paper
- cellophane tape
- soda straws
- felt-tip marker
- meterstick or tape measure
- penny or other coin
- six-sided die

Before You Begin

Natural selection occurs when organisms that have certain **traits** survive to reproduce more than organisms that lack those traits do. A population evolves when individuals with different **genotypes** survive or reproduce at different rates. In this lab, you will model the selection of favorable traits in a new generation by using a paper model of a bird—the fictitious Egyptian origami bird (*Avis papyrus*), which lives in dry regions of North Africa. Assume that only birds that can successfully fly the long distances between water sources will live long enough to breed successfully.

1. Write a definition for each boldface term in the preceding paragraph. Use a separate sheet of paper. **Answers appear in the Teacher's Notes.**

2. You will be using the data table provided to record your data.

3. Based on the objectives for this lab, write a question you would like to explore about the process of selection.

   Answers will vary. For example: Are the effects of natural selection obvious

   in only a few generations?

Modeling Natural Selection *continued*

Procedure

PART A: PARENTAL GENERATION

1. Cut two strips of paper, 2 × 20 cm each. Make a loop with one strip of paper, letting the paper overlap by 1 cm, and tape the loop closed. Repeat for the other strip.

2. Tape one loop 3 cm from each end of the straw, as shown. Mark the front end of the bird with a felt-tip marker. This bird represents the parental generation.

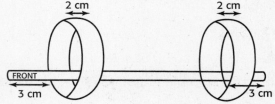

3. Test how far your parent bird can fly by releasing it with a gentle overhand pitch. Test the bird twice. Record the bird's average flight distance in the data table on the next page. **Answers in table will vary.**

PART B: FIRST (F_1) GENERATION

4. Each origami bird lays a clutch of three eggs. Assume that one of the chicks is a clone of the parent. Use the parent to represent this chick in step 6.

5. Make two more chicks. Assume that these chicks have mutations. Follow Steps A–C below for each chick to determine the effects of its mutation.

   Step A Flip a coin to determine which end is affected by a mutation.

   Heads = anterior (front)

   Tails = posterior (back)

   Step B Throw a die to determine how the mutation affects the wing.

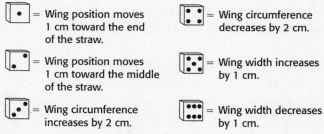

   Step C A mutation is lethal if it causes a wing to fall off the straw or a wing with a circumference smaller than that of the straw. If you get a lethal mutation, disregard it and produce another chick.

6. Record the mutations and the wing dimensions of each offspring.

7. Test each bird twice by releasing it with a gentle overhand pitch. Release the birds as uniformly as possible. Record the distance each bird flies. The most successful bird is the one that flies the farthest.

Modeling Natural Selection *continued*

Data Table

Bird	Coin flip (H or T)	Die throw (1–6)	Anterior wing (cm)			Posterior wing (cm)			Average distance flown (m)
			Width	Circum.	Distance from front	Width	Circum.	Distance from back	
Parent	NA	NA	2	19	3	2	19	3	
Generation 1									
Chick 1									
Chick 2									
Chick 3									
Generation 2									
Chick 1									
Chick 2									
Chick 3									
Generation 3									
Chick 1									
Chick 2									
Chick 3									
Generation 4									
Chick 1									
Chick 2									
Chick 3									
Generation 5									
Chick 1									
Chick 2									
Chick 3									
Generation 6									
Chick 1									
Chick 2									
Chick 3									
Generation 7									
Chick 1									
Chick 2									
Chick 3									
Generation 8									
Chick 1									
Chick 2									
Chick 3									
Generation 9									
Chick 1									
Chick 2									
Chick 3									

PART C: SUBSEQUENT GENERATIONS

8. Assume that the most successful bird in the previous generation is the sole parent of the next generation. Repeat steps 4–7 using this bird.

9. Continue to breed, test, and record data for eight more generations.

PART D: CLEANUP AND DISPOSAL

10. Dispose of paper scraps in the designated waste container.

11. Clean up your work area and all lab equipment. Return lab equipment to its proper place. Wash your hands thoroughly before you leave the lab and after you finish all work.

Analyze and Conclude

1. Analyzing Results Did the birds you made by modeling natural selection fly farther than the first bird you made?

Most students should answer "yes" as the best-flying birds are selected as

the sole parents of the next generation.

2. Inferring Conclusions How might this lab help explain the variety of species of Galápagos finches?

This lab demonstrates that organisms can change significantly over only a

few generations. The lab therefore shows how isolated populations could

diverge to the point that they constitute different species, as happened to

finches on the Galápagos Islands.

3. Further Inquiry Write another question about natural selection that could be explored with another investigation.

Answers will vary. For example: Does natural selection act on one trait at a

time, or can selection pressure affect the evolution of several traits at once?

_____

_____

Making a Dichotomous Key

Teacher Notes

TIME REQUIRED 50 minutes

SKILLS ACQUIRED

Communicating
Designing Experiments
Identifying/Recognizing Patterns
Organizing and Analyzing Data

RATING

Easy ◄—— 1 —— 2 —— 3 —— 4 ——► Hard

Teacher Prep–1
Student Setup–2
Concept Level–2
Cleanup–1

SCIENTIFIC METHODS

In this lab, students will:
Make Observations
Ask Questions
Draw Conclusions
Communicate the Results

MATERIALS

Items commonly found in a classroom can be used or the teacher can ask
students to bring in an assortment of items from home.

SAFETY CAUTIONS

Have students wash hands thoroughly before leaving the lab.

TIPS AND TRICKS

Having students make dichotomous keys allows them to practice observing,
recording, and organizing data. Lead students through a discussion of how to
use the process of elimination when working with a dichotomous key.

You may want to have students complete parts A and B on two different days.
Part A will take about 10 minutes to complete. Part B will require about 40 minutes. Encourage students to select objects that can be easily distinguished from
one another. Emphasize that a dichotomous key includes pairs of descriptions
that lead to the identification of an object.

Making a Dichotomous Key

SKILLS

- Identifying and comparing
- Organizing data

OBJECTIVES

- **Identify** objects using dichotomous keys.
- **Design** a dichotomous key for a group of objects.

MATERIALS

- 6 to 10 objects found in the classroom
  (e.g., shoes, books, writing instruments)
- stick-on labels
- pencil

Before You Begin

One way to identify an unknown organism is to use an **identification key,** which contains the major characteristics of groups of organisms. A **dichotomous key** is an identification key that contains pairs of contrasting descriptions. After each description, a key either directs the user to another pair of descriptions or identifies an object. In this lab, you will design and use a dichotomous key. A dichotomous key can be written for any group of objects.

1. Write a definition for each boldface term in the paragraph above.

   identification key—a tool that can be used to identify an unknown organism

   or object; dichotomous key—a key that uses pairs of contrasting descriptive

   statements to lead to the identification of an organism or some other object

   _____

   _____

2. Based on the objectives for this lab, write a question you would like to explore about making or using a dichotomous key.

   Answers will vary. For example: How do you use a dichotomous key?

   _____

   _____

Making a Dichotomous Key *continued*

Procedure
PART A: USING A DICHOTOMOUS KEY

1. Use the **Key to Forest Trees** to identify the tree that produced each of the leaves shown. Identify one leaf at a time. Always start with the first pair of statements (**1***a* and **1***b*). Follow the direction beside the statement that describes the leaf. Proceed through the key until you get to the name of a tree.

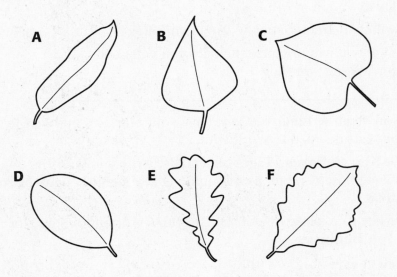

Key to Forest Trees	
1*a* Leaf edge has no teeth, waves, or lobes	go to **2**
1*b* Leaf edge has teeth, waves, or lobes	go to **3**
2*a* Leaf has a bristle at its tip	**shingle oak**
2*b* Leaf has no bristle at its tip	go to **4**
3*a* Leaf edge is toothed	**Lombardy poplar**
3*b* Leaf edge has waves or lobes	go to **5**
4*a* Leaf is heart-shaped	**red bud**
4*b* Leaf is not heart-shaped	**live oak**
5*a* Leaf edge has lobes	**English oak**
5*b* Leaf edge has waves	**chestnut oak**

PART B: DESIGN A DICHOTOMOUS KEY

2. Work with the members of your lab group to design a dichotomous key using the materials listed for this lab.

You Choose

As you design your key, decide the following:

    a. what question you will explore

    b. what objects your key will identify

    c. how you will label personal property

    d. what distinguishing characteristics the objects have

    e. which characteristics to use in your key

    f. how you will organize the data you will need for writing your key

3. Before you begin writing your key, have your teacher approve the objects your group has decided to work with.

4. Using the **Key to Forest Trees** as a guide, write a key for the objects your group selected. Remember, a dichotomous key includes pairs of contrasting descriptions.

5. Use your key to explore one of the questions written for step 2 of **Before You Begin.**

Answers will vary.

_____

_____

6. After each group has completed step 5, exchange keys and the objects they identify with another group. Use the key you receive to identify the objects. If the key does not work, return it to the group so corrections can be made.

PART C: CLEANUP

7. Clean up your work area and all lab equipment. Return lab equipment to its proper place. Wash your hands thoroughly before you leave the lab and after you finish all work.

Analyze and Conclude

1. Drawing Conclusions What tree produced each of the leaves shown in this lab?

A. shingle oak; B. Lombardy poplar; C. red bud; D. live oak; E. English oak;

and F. chestnut oak

_____

_____

Making a Dichotomous Key *continued*

2. Forming Hypotheses What other characteristics might be used to identify leaves with a dichotomous key?

Other leaf characteristics might include whether the leaf is compound or

simple, whether the leaf is needlelike, the arrangement of leaves on the

branches, and the leaf's vein pattern.

_____

3. Analyzing Methods How was the key your group designed dichotomous?

The key was dichotomous because the descriptive statements were written

as pairs of contrasting statements.

_____

4. Evaluating Results Were you able to use another group's key to identify the objects for which it was written? If not, describe the problems you encountered.

Answers will vary. Students might have discovered that some of the opposing

statements were not contrasting, that some of the objects were not correctly

described, or that the key did not include a sufficient number of descriptions

to distinguish all the objects from one another.

_____

5. Analyzing Methods Does a dichotomous key begin with general descriptions and then proceed to more specific descriptions or vice versa? Explain your answer, giving an example from your key.

Dichotomous keys proceed from general characteristics to specific character-

istics. Examples from keys will vary but should reflect this gradation.

_____

6. Further Inquiry Write a new question about making or using keys that could be explored with another investigation.

Answers will vary. For example: Can the members of any kind of group be

identified using a dichotomous key?

_____

Observing How Natural Selection Affects a Population

Teacher Notes

TIME REQUIRED Two 40-minute periods

SKILLS ACQUIRED

Collecting Data

Designing Experiments

Experimenting

Organizing and Analyzing Data

RATING

Teacher Prep–1

Easy ← 1　2　3　4 → Hard

Student Setup–1

Concept Level–2

Cleanup–1

SCIENTIFIC METHODS

In this lab, students will:

Ask Questions

Test the Hypothesis

Analyze the Results

Draw Conclusions

MATERIALS

Materials for this lab can be ordered from WARD'S. Use the Lab Materials QuickList Software on the **One-Stop Planner CD-ROM** for catalog numbers and to create a customized list of materials for this lab. Provide enough metric rulers so each lab group has one. Standard graph paper will be sufficient for the graph. The balance should be sensitive enough to measure 0.1 g. Green beans or snow peas can be obtained from a local grocery store.

SAFETY CAUTIONS

Review all safety symbols with students before beginning the lab. Check to make sure no students are allergic to the food item used.

TIPS AND TRICKS

• Encourage students to read the lab and complete the Before You Begin section before coming to class on the day of the lab.

• During the first lab period, have students make the measurements and record their data. During the second period, have students graph their data and answer the questions.

- Pooling the data will increase the size of the populations studied, thus making students' statistical analyses more valid. To compile class data, record each group's data on the board and have students that measured a similar trait pool the data.

- Make sure students choose a sufficient number of beans or peas to measure so that their data will be more meaningful. Students might choose to measure length of beans or peas, number of seeds in beans or peas, etc.

ANSWERS TO BEFORE YOU BEGIN

1. *natural selection*—the process by which populations change in response to their environment, as better adapted individuals leave more offspring; *variation*—a difference in traits among individuals in a population; *population*—a group of individuals that belong to the same species, live in the same area, and breed with others in the group; *mean*—a mathematical average; *median*—the midpoint in a series of values; *mode*—a value that occurs most frequently; *range*—the difference between the largest and smallest values; *frequency distribution curve*—a type of graph that maps the number of occurrences of each variation of a trait; *stabilizing selection*—a type of selection that increases the proportion of individuals with the average trait; *directional selection*—a type of selection that increases one extreme form of a trait over another form.

Name _____ Class _____ Date _____

Observing How Natural Selection Affects a Population

SKILLS

- Using scientific methods
- Collecting, graphing, and analyzing data

OBJECTIVES

- **Measure** and collect data for a trait in a population.
- **Graph** a frequency distribution curve of your data.
- **Analyze** your data by determining its mean, median, mode, and range.
- **Predict** how natural selection can affect the variation in a population.

MATERIALS

- metric ruler
- graph paper (optional)
- green beans or snow peas
- calculator
- balance

Before You Begin

Natural selection can occur when there is **variation** in a **population**. You can analyze the variation in certain traits of a population by determining the mean, median, mode, and range of the data collected on several individuals. The **mean** is the sum of all data values divided by the number of values. The **median** is the midpoint in a series of values. The **mode** is the most frequently occurring value. The **range** is the difference between the largest and smallest values. The variation in a characteristic can be visualized with a **frequency distribution curve**. Two kinds of natural selection—**stabilizing selection** and **directional selection**—can influence the frequency and distribution of traits in a population. This changes the shape of a frequency distribution curve. In this lab, you will investigate variation in fruits and seeds.

1. Write a definition for each boldface term in the paragraph above. Use a separate sheet of paper. **Answers appear in the Teacher's Notes.**

2. Based on the objectives for this lab, write a question you would like to explore about variation in green beans or snow peas.

   <u>**Answers will vary. For example: How much variation is there in the length of**</u>

   <u>**green beans?**</u>

| **Observing How Natural Selection Affects a Population** *continued*

Procedure
PART A: DESIGN AN EXPERIMENT

1. Work with the members of your lab group to explore one of the questions written for step 2 of **Before You Begin.** To explore the question, design an experiment that uses the materials listed for this lab. **For Example: Measure the lengths or the weights of 20 green beans or snow peas.**

> **You Choose**
>
> As you design your experiment, decide the following:
>
> **a.** what question you will explore
>
> **b.** what hypothesis you will test
>
> **c.** which trait (length, color, weight, etc.) you will measure
>
> **d.** how you will measure the trait
>
> **e.** how many members of the population you will measure (keep in mind that the more data you gather, the more revealing your frequency distribution curve will be)
>
> **f.** what data you will record in your data table

2. Write a procedure for your experiment. Make a list of all the safety precautions you will take. Have your teacher approve your procedure and safety precautions before you begin the experiment.

Sample answer: Organize the measurement values from the lowest to the

highest value, and round to the nearest whole number. Beneath each value,

write the number of green beans of that length. On the graph paper, make a

graph showing the distribution curve.

3. Conduct your experiment. **Students should calculate the mean, median, and range for each type of measurement.**

PART B: CLEANUP AND DISPOSAL

4. Dispose of seeds in the designated waste containers. Do not put lab materials in the trash unless your teacher tells you to do so.

5. Clean up your work area and all lab equipment. Return lab equipment to its proper place. Wash your hands thoroughly before you leave the lab and after you finish all work.

Observing How Natural Selection Affects a Population *continued*

Analyze and Conclude

1. Summarizing Results Make a frequency distribution curve of your data. Plot the trait you measured on the *x*-axis (horizontal axis) and the number of times that trait occurred in your population on the *y*-axis (vertical axis).
Sample graph: Most will approximate a normal distribution.

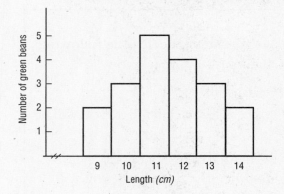

2. Calculating Determine the mean, median, mode, and range of the data for the trait you studied.

Answers will vary.

_____

_____

3. Analyzing Results How does the mean differ from the mode in your population?

Most means will be similar to modes. _____

_____

4. Drawing Conclusions What type of selection appears to have produced the type of variation observed in your experiment?

Most will show stabilizing selection. _____

Observing How Natural Selection Affects a Population *continued*

5. **Evaluating Data** The graph below shows the distribution of wing length in a population of birds on an island. Notice that the mean and the mode are quite different. Is the mean always useful in describing traits in a population? Explain.

Distribution of Wing Length

No. In the example given, the mode in the population is very different

from the mean. Thus, the mean is not always useful in describing traits

in a population.

_____

_____

6. **Forming Hypotheses** What type of selection (stabilizing or directional) would be indicated if the mean of a trait you measured shifted, over time, to the right of a frequency distribution graph?

directional selection

7. **Further Inquiry** Write a new question about variation in populations that could be explored in another investigation.

For example: How does the size of a population of snow peas that has been

grown in hot conditions compare to a population of snow peas grown in

normal weather conditions?

_____

Surveying Kingdom Diversity

Teacher Notes

TIME REQUIRED 40–50 minutes

SKILLS ACQUIRED

Classifying
Collecting Data
Identifying/Recognizing Patterns
Inferring
Organizing and Analyzing Data

RATING

Easy ⟵ 1   2   3   4 ⟶ Hard

Teacher Prep–2
Student Setup–2
Concept Level–2
Cleanup–2

SCIENTIFIC METHODS

In this lab, students will:
Make Observations
Analyze the Results
Draw Conclusions

MATERIALS

Materials for this lab can be ordered from WARD'S. Use the Lab Materials QuickList Software on the **One-Stop Planner CD-ROM** for catalog numbers and to create a customized list of materials for this lab.

SAFETY CAUTIONS

Caution students to use care when observing the preserved specimens. The preservative can leak if the jars are tilted. Tell students to wash their hands if they come into contact with the preservative. Have students wear safety goggles and a lab apron. Remind students about microscope procedures when they use oil-immersion lenses. Remind students to wash their hands before they leave the laboratory.

TIPS AND TRICKS

Provide students with preserved, mounted, or dried samples of archaebacteria, an *Anabaena* species, a protozoan, algae, fungi (mold and yeast), plant tissues, and animals. Set up the specimens at different stations and have students circulate among the stations. Have students make their data table and an answer sheet before class. This will make it easier for students to answer the questions, since not all students will start with Station 1. Students should work in groups of two to four and move clockwise from station to station.

ANSWERS TO BEFORE YOU BEGIN

1. *kingdom*—the largest taxonomic group; *prokaryotic cell*—a cell that has no organized nucleus or organelles; *eukaryotic cell*—a cell that has a nucleus and several different types of organelles; *unicellular organism*—an organism that consists of only one cell; *colonial organism*—a group of cells that are permanently associated but do not communicate with one another; *multicellular organism*—an organism made of many cells that are permanently associated with one another; *tissue*—a group of cells that work together to perform a certain function; *organ*—a specialized structure with a specific function; *organ system*—a group of organs that work together to perform a function; *autotroph*—an organism that obtains energy and makes organic compounds from inorganic materials; *heterotroph*—an organism that obtains energy and organic compounds by eating other organisms

Surveying Kingdom Diversity

SKILLS

- Using a microscope
- Comparing

OBJECTIVES

- **Observe** representatives of each of the six kingdoms.
- **Compare** and **contrast** the organisms within a kingdom.
- **Analyze** the similarities and differences among the six kingdoms.

MATERIALS

- specimens from each of the six kingdoms
- compound microscopes
- hand lenses or stereomicroscopes

SAFETY

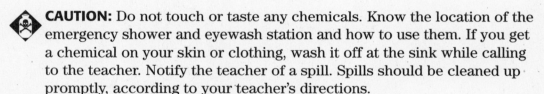

CAUTION: Always wear safety goggles and a lab apron to protect your eyes and clothing.

CAUTION: Do not touch or taste any chemicals. Know the location of the emergency shower and eyewash station and how to use them. If you get a chemical on your skin or clothing, wash it off at the sink while calling to the teacher. Notify the teacher of a spill. Spills should be cleaned up promptly, according to your teacher's directions.

CAUTION: Glassware is fragile. Notify the teacher of broken glass or cuts. Do not clean up broken glass or spills with broken glass unless the teacher tells you to do so.

Before You Begin

Many biologists classify living things into six **kingdoms.** The organisms in a kingdom have fundamental characteristics in common. For example, the organisms of two kingdoms are made of **prokaryotic cells,** while the organisms in the other four kingdoms are made of **eukaryotic cells.** Some kingdoms contain only **unicellular** or **colonial** organisms, while others contain only **multicellular** organisms. In this lab, you will examine representatives of six kingdoms of organisms. You will see that each kingdom is distinct from the others.

1. Write a definition for each boldface term in the previous paragraph and for each of the following terms: tissue, organ, organ system, autotroph, heterotroph. Use a separate sheet of paper. **Answers appear in the Teacher's Notes.**

Surveying Kingdom Diversity *continued*

2. You will be using the data table provided to record your data.

3. Based on the objectives for this lab, write a question you would like to explore about the kingdoms of organisms.

Answers will vary. For example: How do organisms in the six kingdoms

differ?

Procedure

PART A: CONDUCTING A SURVEY

1. Put on safety goggles and a lab apron.

2. Visit the station for each kingdom and examine the specimens there. Answer the questions, and record observations in your data table.

Data Table				
Kingdom name	**Type of cells**	**Level of organization**	**Other characteristics**	**Examples**

3. Archaebacteria Examine the prepared slides.

a. What does a microscope reveal about the structure of archaebacteria?

Archaebacteria are microscopic and prokaryotic.

b. How do these organisms get energy for life processes?

They can be either heterotrophic or autotrophic.

4. Eubacteria Examine the prepared slides.

a. What does a microscope reveal about the structure of eubacteria?

Eubacteria are microscopic, prokaryotic, and many different shapes.

b. Would you consider *Anabaena* to be unicellular or multicellular? Explain.

***Anabaena* appears to be multicellular. It is made of chains of cells.**

c. How does *Anabaena* appear to obtain energy for life processes? Explain.

Anabaena appears to be an autotroph that obtains energy through

photosynthesis.

5. Protists Examine the prepared slides.

a. What does a microscope reveal about the structure of protozoans?

Protozoans are unicellular and much larger than bacteria. They are

eukaryotic, meaning that they have a nucleus and several types of

organelles.

b. How do protozoans appear to obtain energy for life processes? Explain.

Protozoans appear to be heterotrophic since they are not green and

therefore do not contain chlorophyll.

c. Are the algae unicellular or multicellular? Explain.

Some algae are unicellular, and some are multicellular.

d. How do algae differ from protozoans?

Algal cells have cell walls and pigments, such as chlorophyll, which enable

them to be photosynthetic autotrophs.

6. Fungi Examine the specimens.

a. Are fungi unicellular or multicellular? Explain.

Some fungi are unicellular, but most are multicellular.

b. What does a microscope reveal about the structure of fungi?

Fungi are made of strands of undifferentiated cells. These strands are

called hyphae.

c. How do the fungi appear to obtain energy for life processes? Explain.

Fungi appear to be heterotrophic since they are not green and cannot

photosynthesize.

| **Surveying Kingdom Diversity** *continued* |

7. Plants Examine the specimens.

a. What is the most striking characteristic shared by these plants?

Answers will vary. For example, students may note that all of the plants

are green.

b. What does a microscope reveal about the structure of plants?

Plants are made of several types of specialized cells that are arranged in

tissues and organs.

8. Animals Examine the specimens.

a. What is the most striking characteristic shared by these animals?

Answers will vary. For example, students may note that all of the animals

have an organized structure with specialized parts.

b. What is the most striking difference among these animals?

Answers will vary. For example, students may describe differences in size,

shape, or symmetry.

PART B: CLEANUP AND DISPOSAL

9. Dispose of broken glass and solutions in the designated waste containers. Do not pour chemicals down the drain or put lab materials in the trash unless your teacher tells you to do so.

10. Wash your hands thoroughly before you leave the lab and after you finish all work.

Analyze and Conclude

1. Summarizing Data What are the main differences observed among the six kingdoms?

Answers will vary. Students should include differences in cell structure, body

organization, cell specialization, and method of getting food.

_____

_____

2. Recognizing Patterns How does the size of prokaryotic cells compare with the cell size in the other kingdoms?

Cells of archaebacteria and eubacteria are much smaller than the cells of

organisms in the other four kingdoms.

Surveying Kingdom Diversity *continued*

3. Analyzing Methods How did you determine the cell type for each kingdom?

The cell type (prokaryotic or eukaryotic) is determined by microscopic

inspection for a nucleus and organelles.

4. Inferring Conclusions Which kingdom exhibits the most diversity?

Protista is the most diverse, although that may not be evident from the

specimens examined.

5. Further Inquiry Write a new question about the kingdoms of life that could be explored with another investigation.

Answers will vary. For example: How does the method of reproduction differ

among the kingdoms?

Modeling Ecosystem Change over Time

Teacher Notes

TIME REQUIRED About 20 minutes on day 1 and about 10 minutes each day thereafter over a period of a few weeks.

SKILLS ACQUIRED
Collecting Data
Constructing Models
Organizing and Analyzing Data

RATING

Easy ←——1——2——3——4——→ Hard

Teacher Prep–1
Student Setup–2
Concept Level–2
Cleanup–1

SCIENTIFIC METHODS

In this lab, students will:
Make Observations
Test the Hypothesis
Analyze the Results

MATERIALS

Materials for this lab can be ordered from WARD'S. Use the Lab Materials QuickList Software on the **One-Stop Planner CD-ROM** for catalog numbers and to create a customized list of materials for this lab. Have students bring clear plastic 2- or 3-L soda bottles from home. Soil can be collected from around the school, brought from home, or purchased from a garden center. You may also be able to find earthworms and crickets in the local environment or at bait shops.

SAFETY CAUTIONS

Review all safety symbols with students before beginning the lab. Warn students to take care when handling insects and other small animals. Small animals are easily harmed, and some are capable of biting when disturbed.

TIPS AND TRICKS

Remind students that the ecosystems are dependent on humans for care. They should not be permitted to be overheated or become too cold. If water evaporates from the ecosystem, it should be replenished. Recording the number of organisms may be tricky in some cases and estimates may be required.

ANSWERS TO BEFORE YOU BEGIN

1. *ecosystem*—an ecological system encompassing a community and its abiotic factors; *foodweb*—a network of feeding relationships in an ecosystem; *closed ecosystem*—an ecosystem that does not exchange materials outside of itself; *producer*—organisms that first capture energy; *decomposer*—organisms that decompose dead organic material; *consumer*—organisms that consume producers; *herbivore*—organisms that eat plants orother primary producers; *carnivore*—organisms that are secondary consumers; *trophic level*—ecosystem level based on the organism's source of energy.

Exploration Lab

Modeling Ecosystem Change over Time

SKILLS

- Using scientific methods
- Modeling
- Observing

OBJECTIVES

- **Construct** a model ecosystem.
- **Observe** the interactions of organisms in a model ecosystem.
- **Predict** how the number of each species in a model ecosystem will change over time.
- **Compare** a model ecosystem with a natural ecosystem.

MATERIALS

- coarse sand or pea gravel
- large glass jar with a lid or terrarium
- soil
- pinch of grass seeds
- pinch of clover seeds
- rolled oats
- mung bean seeds
- earthworms
- isopods (pill bugs)
- mealworms (beetle larva)
- crickets

Before You Begin

Organisms in an **ecosystem** interact with each other and with their environment. One of the interactions that occurs among the organisms in an ecosystem is feeding. A **food web** describes the feeding relationships among the organisms in an ecosystem. In this lab, you will model a natural ecosystem by building a **closed ecosystem** in a bottle or a jar. You will then observe the interactions of the organisms in the ecosystem and note any changes that occur over time.

1. Write a definition for each boldface term in the paragraph above and for each of the following terms: producer, decomposer, consumer, herbivore, carnivore, trophic level. Use a separate sheet of paper. **Answers appear in the Teacher's Notes.**

Modeling Ecosystem Change over Time *continued*

2. Based on the objectives for this lab, write a question you would like to explore about ecosystems.

<u>Answers will vary. For example, what are the effects of continuous exposure</u>

<u>to bright light on the ecosystem?</u>

Procedure
PART A: BUILDING AN ECOSYSTEM IN A JAR

1. Place 2 in. of sand or pea gravel in the bottom of a large, clean glass jar with a lid. **CAUTION: Glassware is fragile. Notify your teacher promptly of any broken glass or cuts. Do not clean up broken glass or spills with broken glass unless your teacher tells you to do so.** Cover the gravel with 2 in. of soil.

2. Sprinkle the seeds of two or three types of small plants, such as grasses and clovers, on the surface of the soil. Put a lid on the jar, and place it in indirect sunlight. Let the jar remain undisturbed for a week.

3. After one week, place a handful of rolled oats in the jar. Place the mealworms in the oats, and then place the other animals into the jar and replace the lid. Place the lid on the jar loosely to enable air entry.

You Choose

As you design your experiment, decide the following:

    a. what question you will explore

    b. what hypothesis you will test

    c. how you will plant the seeds

    d. where you will place the ecosystem for one week so that it remains undisturbed and in indirect sunlight

    e. how often you will add water to the ecosystem after the first week

    f. how many of each organism you will use

    g. what data you will record in your data table

PART B: DESIGN AN EXPERIMENT

4. Work with the members of your lab group to explore one of the questions written for step 2 of **Before You Begin.** To explore the question, design an experiment that uses the materials listed for this lab.

5. Write a procedure for your experiment. Make a list of all the safety precautions you will take. Have your teacher approve your procedure and safety precautions before you begin the experiment.

6. Set up your group's experiment. Conduct your experiment for at least 14 days.

PART C: CLEANUP AND DISPOSAL

7. Dispose of solutions, broken glass, and other materials in the designated waste containers. Do not put lab materials in the trash unless your teacher tells you to do so.

8. Clean up your work area and all lab equipment. Return lab equipment to its proper place. Wash your hands thoroughly before you leave the lab and after you finish all work.

Analyze and Conclude

1. **Summarizing Results** Make graphs showing how the number of individuals of each species in your ecosystem changed over time. Plot time on the x-axis and the number of organisms on the y-axis.

Answers will vary. Students should make one graph for each species

observed or use different colors to indicate each species.

2. **Analyzing Results** How did your results compare with your hypothesis? Explain any differences.

Answers will vary.

3. **Inferring Conclusions** Construct a food web for the ecosystem you observed.

Answers will vary. All plants are producers (primary trophic level);

earthworms feed on dead plant material in the soil; crickets feed on plants;

mealworms (beetle larvae) feed on plants; isopods (pill bugs) eat wood.

4. **Recognizing Relationships** Does your model ecosystem resemble a natural ecosystem? Explain.

Yes and no. Natural ecosystems and the model ecosystem both contain organ-

isms at several trophic levels, have living and nonliving components, and

depend on the sun for energy. However, the model ecosystem is less diverse,

much younger, and has more definite boundaries than a natural ecosystem.

5. Analyzing Methods How might you have built your model ecosystem differently to better represent a natural ecosystem?

Answers will vary.

_____

_____

_____

6. Evaluating Methods Was your model ecosystem truly a "closed ecosystem"? List your model's strengths and weaknesses as a closed ecosystem.

No, strengths are that the organisms in the model ecosystem did not leave

the ecosystem and that other organisms could not enter from the outside.

Weaknesses are that water and air probably had to be added to maintain a

healthy ecosystem.

_____

_____

7. Further Inquiry Write a new question about ecosystems that you could explore with another investigation.

Answers will vary. For example: What are the effects of certain abiotic

factors, such as temperature, light, and moisture, on the organisms in an

ecosystem?

_____

_____

Observing How Brine Shrimp Select a Habitat

Teacher Notes

TIME REQUIRED Two lab -periods of 40–50 minutes each

SKILLS ACQUIRED

Observing
Designing Experiments
Collecting Data
Communicating

RATING

Easy ←—— 1   2   3   4 ——→ Hard

Teacher Prep–1
Student Setup–2
Concept Level–3
Cleanup–2

SCIENTIFIC METHODS

In this lab, students will:

Make Observations	Analyze the Results
Ask Questions	Draw Conclusions
Test the Hypothesis	

MATERIALS

Materials for this lab can be ordered from WARD'S. Use the Lab Materials QuickList Software on the **One-Stop Planner CD-ROM** for catalog numbers and to create a customized list of materials for this lab.

SAFETY CAUTIONS

• Make sure students do not use very hot water.

• Students should not put their hands near their face while handling the brine shrimp.

• Tell students to wash their hands thoroughly after concluding the experiment.

TIPS AND TRICKS

Students should begin by observing the movement of brine shrimp through water. Tell students to allow the shrimp enough time to distribute themselves in response to environmental factors. This activity works well with groups of four students. When tubing is divided as directed, it allows for space taken up by stoppers and forms equal quarters. Students may count shrimp by looking in the Petri dish or in the pipet. Students might hold the pipet up to a light for better visibility. The Procedure section provides students with one method for randomly dividing brine shrimp into groups that can be experimentally manipulated in the tubing or in the Petri dishes.

Skills Practice Lab

Observing How Brine Shrimp Select a Habitat

SKILLS

- Using scientific methods
- Collecting, organizing, and graphing data

OBJECTIVES

- **Observe** the behavior of brine shrimp.
- **Assess** the effect of environmental variables on habitat selection by brine shrimp.

MATERIALS

- clear, flexible plastic tubing
- metric ruler
- marking pen
- corks to fit tubing
- brine shrimp culture
- screw clamps
- test tubes with stoppers and test-tube rack
- pipet
- Petri dish
- Detain™ or methyl cellulose
- aluminum foil
- calculator
- fluorescent lamp or grow light
- funnel
- graduated cylinder or beaker
- hot-water bag
- ice bag
- pieces of screen
- tape

Before You Begin

Different organisms are adapted for life in different **habitats.** For example, **brine shrimp** are small crustaceans that live in salt lakes. Given a choice, organisms select habitats that provide the conditions (e.g., temperature, light, pH, salinity) to which they are adapted. In this lab, you will investigate habitat selection by brine shrimp and determine which environmental conditions they prefer.

1. Write a definition for each boldface term in the paragraph above.

habitats—areas where organisms live and to which they are adapted

brine shrimp—small crustaceans that live in salt lakes

2. Based on the objectives for this lab, write a question you would like to explore about habitat selection by brine shrimp.

Answers will vary. For example: Will brine shrimp prefer a warm habitat or a

cool habitat?

| Observing How Brine Shrimp Select a Habitat *continued*

Procedure

PART A: MAKING AND SAMPLING A TEST CHAMBER

1. Divide a piece of plastic tubing into 4 sections by making a mark at 12 cm, 22 cm, and 32 cm from one end. Label the sections *1, 2, 3,* and *4.*

2. Place a cork in one end of the tubing. Then transfer 50 mL of brine shrimp culture to the tubing. Place a cork in the open end of the tubing.

3. When you are ready to count shrimp, divide the tubing into four sections by placing a screw clamp at each mark on the tubing. *While someone holds the corks firmly in place,* first tighten the middle clamp and then the outer clamps.

4. Starting at one end, pour the contents of each section into a test tube labeled with the same number. After you empty a section, loosen the adjacent clamp and fill the next test tube.

5. Stopper one test tube, and invert it gently to distribute the shrimp. Use a pipet to transfer a 1 mL sample of shrimp culture to a Petri dish. Add a few drops of Detain™ to the sample. Count and record the number of live shrimp.

6. Repeat step 5 three more times for the same test tube. Record the average number of shrimp for this test tube.

7. Repeat steps 5 and 6 for each of the remaining test tubes.

PART B: DESIGN AN EXPERIMENT

8. Work with the members of your lab group to explore one of the questions written for step 2 of **Before You Begin.** To explore the question, design an experiment that uses the materials listed for this lab.

You Choose

As you design your experiment, decide the following:

    a. what question you will explore

    b. what hypothesis you will test

    c. how to set up your control

    d. how to expose the brine shrimp to the conditions you chose

    e. how long to expose the brine shrimp to the environmental conditions

    f. how you will set up your data table

---

9. Write a procedure for your group's experiment. Make a list of all the safety precautions you will take. Have your teacher approve your procedure and safety precautions before you begin the experiment.

10. Set up and conduct your group's experiment. Do *not* use water over 70°C, which can burn you. **CAUTION: If you are working with the hot-water bag, handle it carefully. If you are working with a lamp, do not touch the bulb. Light bulbs get very hot and can burn your skin.**

PART C: CLEANUP AND DISPOSAL

11. Dispose of broken glass in the designated waste container. Put brine shrimp in the designated container. Do not pour chemicals down the drain or put lab materials in the trash unless your teacher tells you to do so.

12. Clean up your work area and all lab equipment. Return lab equipment to its proper place. Wash your hands thoroughly before you leave the lab and after you finish all work.

Analyze and Conclude

1. **Summarizing Results** Make a bar graph of your data. Use graph paper to plot the environmental variable on the *x*-axis and the number of shrimp on the *y*-axis.

2. **Analyzing Results** How did the shrimp react to changes in the environment?

   Answers will vary depending on the species of *Artemia* used.

3. **Analyzing Methods** Why was a control necessary?

   The control was necessary to show that the brine shrimp did not inherently

   prefer one part of the tube.

4. **Analyzing Methods** Why was it necessary to take many counts in each test tube (step 6 of Part A)?

   Many counts were taken to make allowances for variations in populations

   and to provide data for calculating an average.

5. **Further Inquiry** Write a new question about brine shrimp that could be explored with another investigation.

   Answers will vary. For example: How do brine shrimp react to water

   movement?

Studying Population Growth

Teacher Notes

TIME REQUIRED 20–30 minutes per day for 4 consecutive days

SKILLS ACQUIRED

Collecting Data
Inferring
Measuring
Organizing and Analyzing Data
Predicting

RATING

Easy ← 1   2   3   4 → Hard

Teacher Prep–1
Student Setup–2
Concept Level–3
Cleanup–2

SCIENTIFIC METHODS

In this lab, students will:
Make Observations
Analyze the Results
Draw Conclusions

MATERIALS

Materials for this lab can be ordered from WARD'S. Use the Lab Materials QuickList Software on the **One-Stop Planner CD-ROM** for catalog numbers and to create a customized list of materials for this lab.

SAFETY CAUTIONS

Caution students to treat all micro-organisms as potential pathogens. Remind students to keep their hands away from their faces as they handle the yeast cultures. Remind students to wash their hands after this lab.

TIPS AND TRICKS

Prepare the yeast population by dissolving 1.0 g of yeast and 1.0 g of sugar in 40 mL of warm water. Remove 1 mL of this solution and dilute with 9 mL of water. Mix well and again remove 1 mL of the solution and dilute with 9 mL of water. Fresh yeast may be prepared up to 4 hours before the investigation. If substituting dried yeast or freeze-dried yeast (available from WARD'S), prepare about a week in advance. Keep the yeast in a warm, dark area for the duration of the investigation. Ruled microscope slides can be purchased from WARD'S. Alternatively, make a transparency copy of a piece of graph paper and then cut the transparency into coverslip-size pieces. Students can use these transparency pieces as their coverslips. Use gloves when preparing methylene blue and avoid creating dust while working. Make a 1.0 percent solution by dissolving 1.0 g of methylene blue in 100 mL of distilled water. Have students work in teams of two. Solutions of yeast and of methylene blue may be rinsed down the drain. Wash thoroughly and air dry all glassware.

ANSWERS TO BEFORE YOU BEGIN

1. *limiting factors*—resources needed for survival obtained from or provided by the environment; *population growth*—increase in population size; *birthrate*—number of births in a population per unit time; *death rate*—number of deaths in a population per unit time

ANSWER TO ANALYZE AND CONCLUDE

2. Sample Data Graph: Estimate of Yeast Population Growth

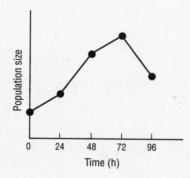

Name _____ Class _____ Date _____

Studying Population Growth

SKILLS

- Using a microscope
- Collecting, graphing, and analyzing data
- Calculating

OBJECTIVES

- **Observe** the growth and decline of a population of yeast cells.
- **Determine** the carrying capacity of a yeast culture.

MATERIALS

- safety goggles
- lab apron
- yeast culture
- (2) 1 mL pipets
- 2 test tubes
- 1% methylene blue solution
- ruled microscope slide (2 × 2 mm)
- coverslip
- compound microscope

SAFETY

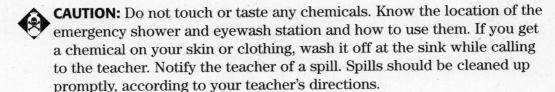

CAUTION: Always wear safety goggles and a lab apron to protect your eyes and clothing.

CAUTION: Do not touch or taste any chemicals. Know the location of the emergency shower and eyewash station and how to use them. If you get a chemical on your skin or clothing, wash it off at the sink while calling to the teacher. Notify the teacher of a spill. Spills should be cleaned up promptly, according to your teacher's directions.

CAUTION: Glassware is fragile. Notify the teacher of broken glass or cuts. Do not clean up broken glass or spills with broken glass unless the teacher tells you to do so.

Before You Begin

Recall that population size is controlled by **limiting factors**—environmental resources such as food, water, oxygen, light, and living space. **Population growth** occurs when a population's **birth rate** is greater than its **death rate**. A decline in population size occurs when a population's death rate surpasses its birth rate. In this lab, you will study the concepts of population growth, decline, and carrying capacity by growing and observing yeast.

1. Write a definition for each boldface term in the preceding paragraph. Use a separate sheet of paper. **Answers appear in the Teacher's Notes.**

2. You will be using the data table provided on the next page to record your data.

3. Based on the objectives for this lab, write a question about population growth that you would like to explore.

   For example: How does the size of a population with limited resources
   _____

   change over time?
   _____

Procedure
PART A: COUNTING YEAST CELLS

1. Put on safety goggles and a lab apron.

2. Transfer 1 mL of a yeast culture to a test tube. Add 2 drops of methylene blue to the tube. **Caution: Methylene blue will stain your skin and clothing.** The methylene blue will remain blue in dead cells but will turn colorless in living cells.

3. Make a wet mount by placing 0.1 mL (one drop) of the yeast and methylene blue mixture on a ruled microscope slide. Cover the slide with a coverslip.

4. Observe the wet mount under the low power of a compound microscope. Notice the squares on the slide. Then switch to the high power. *Note: Adjust the light so that you can clearly see both stained and unstained cells.* Move the slide so that the top left-hand corner of one square is in the center of your field of view. This will be area 1, as shown in the diagram below.

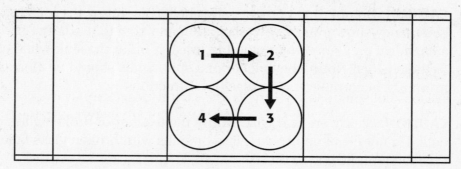

Studying Population Growth *continued*

5. Count the live (unstained) cells and the dead (stained) cells in the four corners of a square using the pattern shown in the diagram. In the data table below, record the numbers of live cells and dead cells in the square.

Answers will vary. _____

Data Table			
Time (hours)	Number of cells per square		Population size (cells/0.1 mL)
	Squares 1–6	Average	
0			
24			
48			
72			
96			

6. Repeat step 5 until you have counted 6 squares on the slide. Complete Part B.

Answers will vary. _____

7. Find the total number of live cells in the 6 squares. Divide this total by 6 to find the average number of live cells per square. Record this number in the data table. Repeat this procedure for dead cells.

Answers will vary. _____

8. Estimate the population of live yeast cells in 1 mL (the amount in the test tube) by multiplying the average number of cells per square by 2,500. Record this number in the data table. Repeat this procedure for dead cells.

Answers will vary. _____

9. Repeat steps 1 through 8 each day for 4 more days.

Answers will vary. _____

PART B: CLEANUP AND DISPOSAL

10. Dispose of solutions and broken glass in the designated waste containers. Do not pour chemicals down the drain or put lab materials in the trash unless your teacher tells you to do so.

11. Clean up your work area and all lab equipment. Return lab equipment to its proper place. Wash your hands thoroughly before you leave the lab and after you finish all work.

Analyze and Conclude

1. Analyzing Methods Why were several areas and squares counted and then averaged each day?

An average was taken to allow for variation within the population.

2. Summarizing Results Use graph paper to graph the changes in the numbers of live yeast cells and dead yeast cells over time. Plot the number of cells in 1 mL of yeast culture on the *y*-axis and the time (in hours) on the *x*-axis. **A sample data graph appears in the Teacher's Notes.**

3. Inferring Conclusions What limiting factors probably caused the yeast population to decline?

A lack of food and lack of space limit the yeast cells. They could also be

poisoned by their own wastes.

4. Further Inquiry Write a new question about population growth that could be explored in another investigation.

Answers will vary. For example: Would the carrying capacity of the yeast's

environment expand if the size of the environment increased?

Staining and Observing Bacteria

Teacher Notes

TIME REQUIRED 45 minutes

SKILLS ACQUIRED

Classifying
Collecting Data
Experimenting
Organizing and Analyzing Data

RATING

Easy ⟵———1———2———3———4———⟶ Hard

Teacher Prep–3
Student Setup–3
Concept Level–2
Cleanup–2

SCIENTIFIC METHODS

In this lab, students will:
Make Observations
Draw Conclusions
Communicate the Results

MATERIALS

Materials for this lab can be ordered from WARD'S. Use the Lab Materials QuickList Software on the **One-Stop Planner CD-ROM** for catalog numbers and to create a customized list of materials for this lab. All cultures should be autoclaved before disposal. Have students place their used swabs in a centrally located container. The swabs may then be incinerated or autoclaved before disposal.

SAFETY CAUTIONS

Make sure students wear safety goggles, lab aprons, and disposable gloves before obtaining their bacterial cultures. Remind students to use methylene blue carefully in step 7 and to keep it away from their faces, skin, and clothing. Be sure that students wipe their tables down with the rubbing alcohol before and after the lab. Be sure that open flames are isolated from the areas where alcohol is being used. Make sure that students dispose of their materials properly and wash their hands before leaving the lab.

TIPS AND TRICKS

There should be sufficient wax pencils (or permanent marking pens) to supply three or four students with one pencil. You may purchase live cultures or lyophilized cultures from a biological supply house such as WARD'S. Purchase *Micrococcus luteus* (coccus), *Spirillum volutans* (spirillum), and *Bacillus megaterium* (bacillus). Standard isopropyl alcohol is sufficient for aseptic-technique washing down of the tables. Provide students with sufficient sterile cotton swabs that are wrapped individually or paired. Buy premixed nutrient broth from a biological supply house such as WARD'S, or prepare culture medium from dehydrated concentrate and sterilize in an autoclave at 120°C at 15 psi for 20 minutes. About one day before the lab, grow the bacteria in nutrient broth at 37°C. Prepare a sufficient amount of culture broth to provide groups of three or four students with about 5 mL of each culture. Transfer about 5 mL of culture to each tube with a sterile pipet. Do not allow students to access your stock cultures. If students' microscopes have an oil immersion lens, you may provide them with immersion oil so that they can view the bacteria at a higher magnification.

Each lab group should be given a set of bacterial cultures. If you allow groups to obtain samples directly from stock cultures, the stocks are likely to become contaminated. Use broth cultures, which will enable students to see chains of cocci.

ANSWERS TO BEFORE YOU BEGIN

1. *prokaryote*—organism that has no nucleus or membrane-enclosed structures; *smear*—microscope slide that has bacterial cells smeared and dried on it; *stain*—a colored substance that makes cells more visible under magnification; *strepto*—prefix used to describe bacteria that form chains of cells; *staphylo*—prefix used to describe bacteria that form clusters of cells; *coccus*—spherical bacterium; *bacillus*—cylindrical- or rod-shaped bacterium; *spirillum*—corkscrew-shaped bacterium.

Name _____ Class _____ Date _____

Skills Practice Lab

Staining and Observing Bacteria

SKILLS

- Using aseptic techniques
- Using a microscope

OBJECTIVES

- **Prepare** and stain wet mounts of bacteria.
- **Identify** different types of bacteria by their shape.

MATERIALS

- wax pencil
- 3 microscope slides
- safety goggles
- lab apron
- disposable gloves
- rubbing alcohol
- paper towels
- 3 culture tubes of bacteria (A, B, and C)
- test-tube rack
- sterile cotton swabs
- Bunsen burner with striker
- microscope slide
- forceps or wooden alligator-type clothespin
- 150 mL beaker
- methylene blue stain in dropper bottle
- compound microscope

SAFETY

 CAUTION: Always wear safety goggles and a lab apron to protect your eyes and clothing.

 CAUTION: Do not touch or taste any chemicals. Know the location of the emergency shower and eyewash station and how to use them. If you get a chemical on your skin or clothing, wash it off at the sink while calling to the teacher. Notify the teacher of a spill. Spills should be cleaned up promptly, according to your teacher's directions.

 CAUTION: Glassware is fragile. Notify the teacher of broken glass or cuts. Do not clean up broken glass or spills with broken glass unless the teacher tells you to do so.

Before You Begin

Like all **prokaryotes,** bacteria are unicellular organisms that sometimes form filaments or loose clusters of cells. They are prepared for viewing by making a **smear,** a slide on which cells have been spread and dried. Treating the cells with a **stain** makes them more visible under magnification. In this lab, you will stain, identify, and compare and contrast different types of bacteria.

1. Write a definition for each boldface term in the paragraph above and for each of the following terms: strepto, staphylo, coccus, bacillus, spirillum. Use a separate sheet of paper. **Answers appear in the Teacher's Notes.**

2. Based on the objectives for this lab, write a question you would like to explore about different kinds of bacteria.

   Answers will vary. For example: What physical differences can we observe in

   different types of bacteria when seen under a compound light microscope?

Procedure

PART A: OBSERVING LIVE BACTERIA

1. Put on safety goggles, a lab apron, and disposable gloves.

2. Use a wax pencil to label three microscope slides *A, B,* and *C.*

3. Use rubbing alcohol and paper towels to clean the surface of your lab table and gloves. Allow the table to air dry. **CAUTION: Alcohol is flammable. Do not use alcohol near an open flame.**

4. Light a Bunsen burner with a striker. **CAUTION: Keep combustibles away from flames. Do not light a Bunsen burner when others in the room are using alcohol.**

5. Beginning with culture A, make a smear of three different bacteria (A, B, and C) as follows. Remove the cap from a culture tube. *Note: Do not place the cap on the table.* Pass the opening of the tube through the flame of a Bunsen burner. Insert a sterile cotton swab into the tube, and lightly touch the tip of the swab to the bacteria in the culture. Pass the opening of the tube through the flame again, and replace the cap. Transfer a small amount of bacteria to the appropriately labeled microscope slide by rubbing the swab on the slide. Dispose of the swab in a proper container. Repeat for cultures B and C.

6. Allow your smears to air dry. Using microscope slide forceps, pick up each slide one at a time and pass it over the flame several times. Let each slide cool.

7. Using microscope slide forceps, place one of your slides across the mouth of a 150 mL beaker half-filled with water. Place 2–3 drops of methylene blue stain on the dried bacteria. *Note: Do not allow the stain to spill into the beaker.* **CAUTION: Methylene blue will stain your skin and clothing.** Let the stain stay on the slide for 2 minutes. Then dip the slide into the water in the beaker several times to rinse it. Blot the slide dry with a paper towel. *Note: Do not rub the slide.*

8. Repeat step 7 for your other two slides.

9. Allow each slide to air dry, and then observe them with a microscope. Make a sketch of a few cells on each slide. Identify the type of bacteria on each slide.

PART B: CLEANUP AND DISPOSAL

10. Dispose of slides, used swabs, solutions, and broken glass in the designated waste containers. Do not pour chemicals down the drain or put lab materials in the trash unless your teacher tells you to do so.

11. Clean up your work area and all lab equipment. *Clean the surface of your lab table with rubbing alcohol.* Return lab equipment to its proper place. Wash your hands thoroughly before you leave the lab and after you finish all work.

Analyze and Conclude

1. **Summarizing Results** Describe the shape and grouping of the cells of each type of bacteria you observed.

All three types of bacterial cells should have been observed. Cocci may form

pairs, groups of four, chains, or clusters. Bacilli usually appear as single

cells, although sometimes they are arranged in pairs or in chains. Spirilla

almost always appear singly, and they differ in length and in the number and

size of their spirals.

2. **Analyzing Methods** Why should the test tube caps from the culture tubes (in Part B) not be placed on the table?

Culture tube caps should not be placed on lab tables because bacteria on the

tables could touch the cap and contaminate the bacterial cultures.

_____

Staining and Observing Bacteria *continued*

3. Evaluating Viewpoints Evaluate the following advice: Always use caution when handling bacteria, even if the bacteria is known to be harmless.

Answers will vary. Students should mention that mistakes in identification,

contamination, or mutations may result in the presence of harmful bacteria.

_____

4. Drawing Conclusions How did you classify the bacteria in cultures A, B, and C—as a coccus, a bacillus, or a spirillum?

Answers will vary according to the type of bacteria placed into the marked

tubes.

5. Further Inquiry Write a new question about bacteria that could be explored with another investigation.

Answers will vary. For example: Are antiseptics equally effective against the

three types of bacteria?

Observing Protistan Responses to Light

Teacher Notes

TIME REQUIRED 45 to 75 minutes

SKILLS ACQUIRED
Collecting Data
Designing Experiments
Experimenting
Organizing and Analyzing Data

RATING
Easy ← 1   2   3   4 → Hard

Teacher Prep–2
Student Setup–2
Concept Level–2
Cleanup–2

SCIENTIFIC METHODS

In this lab, students will:
Make Observations
Ask Questions
Test the Hypothesis
Analyze the Results

MATERIALS

Materials for this lab can be ordered from WARD'S. Use the Lab Materials QuickList Software on the **One-Stop Planner CD-ROM** for catalog numbers and to create a customized list of materials for this lab.

SAFETY CAUTIONS

Discuss all safety symbols and caution statements with students. Protists and Detain™ may be washed down the drain.

TIPS AND TRICKS

This investigation may take more than one class period. To save time, you may wish to make the "sun shades" in advance. (Students should consider that some protists benefit from light and others do not; this might affect the organisms' responses to light.)

ANSWERS TO BEFORE YOU BEGIN

1. *protists*—member of Kingdom Protista

Kingdom Protista—a diverse group of eukaryotic organisms that cannot be classified as animals, plants, or fungi

eukaryotes—organisms that have complex cells with a nucleus enclosed by a membrane

producer—an organism that makes its own food from light energy and inorganic molecules in its environment

consumer—an organism that must consume other organisms to obtain energy

decomposer—an organism that breaks down organic matter

cilia—tightly packed rows of short hairlike structures used for movement

flagellum—a whiplike structure used for movement

pseudopod—a flexible cytoplasmic extension used for movement and obtaining food

Exploration Lab

Observing Protistan Responses to Light

SKILLS

- Using scientific methods
- Using a microscope

OBJECTIVES

- **Identify** several different types of protists.
- **Compare** the structures, methods of locomotion and feeding, and behaviors of several different protists.
- **Relate** a protist's response to light to its method of feeding.

MATERIALS

- Detain™ (protist-slowing agent)
- microscope slides
- plastic pipets with bulbs
- assorted cultures of protists
- toothpicks
- coverslips
- compound microscope
- protist references
- black construction paper
- scissors
- paper punch
- white paper
- sunlit window sill or lamp
- forceps

Before You Begin

Protists belong to the kingdom **Protista**, which is a diverse group of **eukaryotes** that cannot be classified as animals, plants, or fungi. Many protists are unicellular. Among the protists, there are **producers, consumers,** and **decomposers.** In this lab, you will observe live protists and compare their structures, methods of locomotion and feeding, and behaviors.

1. Write a definition for each boldface term in the paragraph above and for each of the following terms: cilia, flagellum, pseudopod. Use a separate sheet of paper. **Answers appear in the Teacher's Notes.**

2. You will be using the data table provided to record your data.

| Observing Protistan Responses to Light *continued*

3. Based on the objectives for this lab, write a question you would like to explore about protists.

Answers will vary. For example: How do *Euglena* and *Volvox* respond to light?

Procedure
PART A: MAKE OBSERVATIONS

1. Caution: Do not touch your face while handling microorganisms. Place a drop of Detain™ on a microscope slide. Add a drop of liquid from the bottom of a mixed culture of protists. Mix the drops with a toothpick. Add a coverslip. View the slide under low power of a microscope. Switch to high power.

2. Use references to identify the protists. Record data for each type of protist.

Sample Data:

<table>
<tr><th colspan="5">Data Table</th></tr>
<tr><th>Protist</th><th>Color</th><th>Method
of locomotion</th><th>Method
of feeding</th><th>Other
observations</th></tr>
<tr><td>*Amoeba*</td><td>Whitish</td><td>Pseudopodia</td><td>Consumer</td><td>Changes shape</td></tr>
<tr><td>*Blepharisma*</td><td>Pink</td><td>Cilia</td><td>Consumer</td><td></td></tr>
<tr><td>*P. caudatum*</td><td>Clear</td><td>Cilia</td><td>Consumer</td><td></td></tr>
<tr><td>*Euglena*</td><td>Green</td><td>Flagellum</td><td>Producer</td><td>Has "eye-spot"</td></tr>
<tr><td>*Stentor*</td><td>Bluish</td><td>Cilia</td><td>Consumer</td><td>Trumpet-shaped</td></tr>
<tr><td>*Volvox*</td><td>Green</td><td>Flagellum</td><td>Producer</td><td>Colonial</td></tr>
</table>

3. Repeat step 1 *without* using Detain™.

4. Punch a hole in a 40 × 20 mm piece of black construction paper that has a slight curl. The paper should be curled along its long axis with the hole in the middle.

5. Place a wet mount of protists on a piece of white paper. Then put the paper and slide on a sunlit window sill or under a table lamp. Position the sun shade on top of the slide so that the hole is in the center of the coverslip.

6. To examine a slide, first view the area in the center of the hole under low power. *Note: Do not disturb the sun shade. Do not switch to high power.* Then have a partner carefully remove the sun shade with forceps while you observe the slide.

Observing Protistan Responses to Light *continued*

PART B: DESIGN AN EXPERIMENT

7. Work with members of your lab group to explore one of the questions written for step 3 of **Before You Begin.** To explore the question, design an experiment that uses the materials listed for this lab.

> **You Choose**
>
> As you design your experiment, decide the following:
>
> **a.** what question you will explore
>
> **b.** what hypothesis you will test
>
> **c.** how long you will expose protists to light
>
> **d.** how many times you will repeat your experiment
>
> **e.** what your control will be
>
> **f.** what data you will record and how you will make your data table

8. Write a procedure for your experiment. Make a list of all the safety precautions you will take. Have your teacher approve your procedure and safety precautions before you begin the experiment.

9. Set up and carry out your experiment.

PART C: CLEANUP AND DISPOSAL

10. Dispose of lab materials and broken glass in the designated waste containers. Put protists in the designated containers. Do not put lab materials in the trash unless your teacher tells you to do so.

11. Clean up your work area and all lab equipment. Return lab equipment to its proper place. Wash your hands thoroughly before you leave the lab and after you finish all work.

Analyze and Conclude

1. Summarizing Results Describe the different types of locomotion you observed in protists, and give examples of each.

Protists may move by using cilia, flagella, or pseudopodia. For example,

Amoeba moves by flowing cytoplasm into a pseudopodium, *Paramecium* uses

cilia beating in waves to propel the cell, *Euglena* uses a whiplike flagellum, and

Volvox uses many flagella.

_____

_____

Observing Protistan Responses to Light *continued*

2. Analyzing Results Identify which protists were affected by light, and describe how they were affected.

Sample answer: *Euglena* **congregated within the area exposed to light. Once**

the shade was removed, they scattered. Euglenoids have an eyespot used to

detect the light needed for photosynthesis.

3. Inferring Conclusions What is the relationship between a protist's response to light and its method of feeding?

Protozoans that depend on light for food were attracted to the light.

4. Further Inquiry Write a new question about protists that could be explored with another investigation.

Answers will vary. For example: Do samples from the middle and upper areas of

the culture jar have different protist populations from those taken at the

bottom?

Observing Yeast and Fermentation

Teacher Notes

TIME REQUIRED One 50-minute class period

SKILLS ACQUIRED
Observing
Measuring
Collecting Data
Analyzing Data

RATING

Easy ←——— 1   2   3   4 ——→ Hard

Teacher Prep–3
Student Setup–2
Concept Level–3
Cleanup–2

SCIENTIFIC METHODS

In this lab, students will:
Make Observations
Analyze the Results
Draw Conclusions
Communicate the Results

MATERIALS

Materials for this lab can be ordered from WARD'S. Use the Lab Materials QuickList Software on the **One-Stop Planner CD-ROM** for catalog numbers and to create a customized list of materials for this lab.

SAFETY CAUTION

Review all safety symbols with students before beginning the lab.

TIPS AND TRICKS

Review the fundamentals of graphing prior to this investigation. Prepare a saturated limewater solution by combining approximately 700 mg of calcium carbonate per 500 ml of distilled water. Allow the solution to sit overnight and filter before using. The limewater should become cloudy as the carbon dioxide gas produced by the yeast flows into the flask. An indicator solution using bromthymol blue instead of limewater may also be also be used to detect the presence of carbon dioxide gas in the flask. This solution can be prepared by dissolving 0.25 g of bromthymol blue powder per 50 ml of distilled water. The solution should be green or blue. If it is yellow, the solution should be slowly titrated with a 0.1% sodium hydroxide solution only until it turns blue or dark green. The solution should turn yellow as the concentration of carbon dioxide gas present in the flask increases.

Name _____ Class _____ Date _____

Observing Yeast and Fermentation

SKILLS

- Observing
- Measuring
- Collecting data
- Analyzing data

OBJECTIVE

- **Observe** the release of energy by yeast during fermentation.

MATERIALS

- 500 mL vacuum bottle
- 10 cm glass tubing
- 2-hole rubber stopper
- 250 mL beaker
- 75 g sucrose
- one package dry yeast
- thermometer
- 50 cm rubber tubing
- 150 mL limewater

SAFETY

  CAUTION: Always wear safety goggles and a lab apron to protect your eyes and clothing.

 CAUTION: Glassware is fragile. Notify the teacher of broken glass or cuts. Do not clean up broken glass or spills with broken glass unless the teacher tells you to do so.

Before You Begin

Sucrose is a disaccharide—a carbohydrate made from two monosaccharides. It is one chemical made by plants to store the sun's energy. Yeast release the energy stored in sucrose in a process called **fermentation.** In this investigation you will have a chance to observe and measure the products of fermentation.

1. Write a definition for the boldface term in the paragraph above.

   Fermentation is the process by which cells release the energy in sugar with-

   out using oxygen.

Observing Yeast and Fermentation *continued*

2. You will be using the data table below to record your data.

Data Table								
Fermentation by Yeast								
Time	**Date**	**Temp.**	**Time**	**Date**	**Temp.**	**Time**	**Date**	**Temp.**
1.			8.			15.		
2.			9.			16.		
3.			10.			17.		
4.			11.			18.		
5.			12.			19.		
6.			13.			20.		
7.			14.			21.		

Procedure

1. Set up your vacuum bottle according to the diagram.

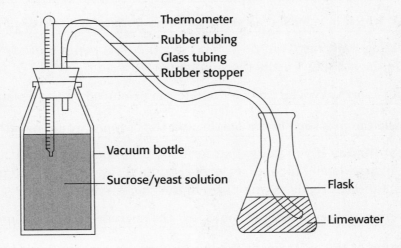

2. Mix 75 g of sucrose in 400 mL of water.

3. When the sucrose has dissolved, add one-half package of fresh yeast and stir.

4. Pour the sucrose-yeast solution into a vacuum bottle until it is approximately three-quarters full.

5. Adjust the thermometer so that it extends down into the sugar-yeast solution.

Observing Yeast and Fermentation *continued*

6. Record the temperature of the solution on the observation chart as soon as possible. Continue to record the temperature as often as possible during the next two days.

Analyze and Conclude

1. Summarizing Results Prepare a graph of your data, illustrating the temperature over time. Complete the graph by drawing a curve through the plotted points.

Students' graphs may vary. They should show that as fermentation

progresses, the temperature in the sucrose-yeast solution increases.

2. Analyzing Data What does the curved line plotted on the graph indicate?

Answers may vary. Students should interpret the curved line as an indication

that fermentation activity increases over time.

3. Drawing Conclusions What can you conclude about the energy contained in sucrose?

Answers may vary. Since the temperature of the sucrose-yeast solution

increases over time, students should conclude that as fermentation occurs,

large amounts of energy are being released. This would indicate that the

sucrose molecules contain a large amount of energy in their molecular

bonds, which is released as the yeast metabolizes the sucrose.

4. Predicting Patterns What do you think would happen if there were only one hole in the stopper for the thermometer?

Answers may vary. The carbon dioxide gas produced by the yeast would

increase the pressure in the bottle, and the stopper would eventually come off.

5. Further Inquiry If you know that fermentation liberates energy and gives off carbon dioxide and alcohol as waste products, how would you prove that fermentation is really taking place in the sugar-yeast solution?

Answers may vary. Students could test the sucrose-yeast solution at the

beginning of the experiment for the presence of carbon dioxide by adding a

small amount to an indicator solution.

Surveying Plant Diversity

Teacher Notes

TIME REQUIRED 45 minutes

SKILLS ACQUIRED
Classifying
Collecting Data
Identifying/Recognizing Patterns
Inferring

RATING

Easy ◄——— 1   2   3   4 ———► Hard

Teacher Prep–1
Student Setup–1
Concept Level–2
Cleanup–1

SCIENTIFIC METHODS

In this lab, students will:
Make Observations
Ask Questions
Test the Hypothesis
Analyze the Results
Draw Conclusions
Communicate the Results

MATERIALS

Materials for this lab can be ordered from WARD'S. Use the Lab Materials QuickList Software on the **One-Stop Planner CD-ROM** for catalog numbers and to create a customized list of materials for this lab.

SAFETY CAUTIONS

Review all safety symbols and caution statements with students. Remind students to be careful with glass containers and prepared slides. Review procedure for cleaning and disposing of broken glass.

TIPS AND TRICKS

Review the four major plant groups before conducting lab. Set up at least two stations for each group of plants. Provide a stereomicroscope or hand lens for stations with mosses and ferns. Provide a compound microscope for stations with prepared slides.

ANSWERS TO BEFORE YOU BEGIN

1. *alternation of generations*—a lifecycle that alternates between a haploid gametophyte stage and a diploid sporophyte stage; *gametophyte*—the gamete-producing haploid stage of a plant's life cycle; *sporophyte*— the spore-producing diploid stage of a plant's life cycle; *sporangium*—structure in which spores are produced; *spore*—haploid asexual reproductive structure that is resistant to environmental stress; *frond*—leaf of a fern; *cone*—reproductive structure of the sporophyte found in certain plants; *flower*—reproductive structure of angiosperm in which the gametophytes- develop; *fruit*—mature plant ovary in angiosperms; adapted for dispersal

Skills Practice Lab

Surveying Plant Diversity

SKILLS

- Observing
- Comparing

OBJECTIVES

- **Identify** similarities and differences among four phyla of living plants.
- **Relate** structural adaptations of plants to their success on land.

MATERIALS

- live or preserved specimens of mosses, ferns, conifers, and flowering plants
- stereomicroscope or hand lens
- compound microscope
- prepared slides of fern gametophytes

Before You Begin

Most plants are complex photosynthetic organisms that live on land. The ancestors of plants lived in water. As plants evolved on land, however, they developed adaptations that made it possible for them to be successful in dry conditions. All plant life cycles are characterized by **alternation of generations,** in which a haploid **gametophyte** stage alternates with a diploid **sporophyte** stage. Distinct differences in the relative sizes and structures of gametophytes and sporophytes are seen among the 12 phyla of living plants. In this lab, you will examine representatives of the four most familiar plant phyla.

1. Write a definition for each boldface term in the paragraph above and for the following terms: sporangium, spore, frond, cone, flower, fruit. Use a separate sheet of paper. **Answers appear in the Teacher's Notes.**

2. You will be using the data table provided to record your data.

3. Based on the objectives for this lab, write a question you would like to explore about the characteristics of plants.

   Answers will vary. For example: What characteristics are shared by all plants?

Procedure

PART A: CONDUCTING A SURVEY

1. Visit the station for each of the plants , and examine the specimens there. Answer the questions, and record observations in your data table.

Data Table			
Phylum name	**Dominant generation**	**Major characteristics**	**Examples**

2. **Mosses** Examine a clump of moss with a stereomicroscope or hand lens. Make a sketch of what you see. Use a separate sheet of paper.

3. **Mosses** Examine a moss gametophyte with a sporophyte attached to it. Draw what you see, and label the parts you recognize. Label each part as haploid or diploid.

   a. Which stage of a moss has rootlike structures?

   gametophyte

   b. Where are the spores of a moss produced?

   sporophyte

4. **Ferns** Examine the sporophyte of a fern, and look for evidence of reproductive structures on the underside of the fronds. Draw what you see. Label a leaf (frond), stem, root, and reproductive structure.

   a. How does water travel through a fern? List observations supporting your answer.

   Water travels through the cells of the vascular system. Veins are visible in

   the leaves.

   b. What kind of reproductive cells are produced by fern fronds?

   spores

5. **Ferns** Examine a slide of a fern gametophyte with a compound microscope. Draw what you see, and label any structures you recognize.

   In general, a fern gametophyte should be relatively heart-shaped.

6. Conifers Draw a part of a branch of one of the conifers at this station. Label a leaf, stem, and cone (if present).

a. Is a branch of a pine tree part of a gametophyte or part of a sporophyte?

sporophyte _____

b. In what part of a conifer would you look to find its reproductive structures?

cones _____

7. Conifers Examine a prepared slide of pine pollen. Draw a few of the grains.

a. What reproductive structure is found within a pollen grain?

sperm _____

b. How does the structure of pine pollen aid in its dispersal by wind?

Appendages help the pollen grains to sail in the wind. _____

8. Angiosperms Draw one of the representative angiosperms at this station. Label a leaf, stem, root, and flower (if present). Indicate the sporophyte and location of gametophytes.

a. Where do angiosperms produce sperm and eggs?

in flowers _____

b. How do the seeds of angiosperms differ from those of gymnosperms?

Angiosperm seeds are produced in fruits; gymnosperm seeds are produced in

cones. _____

9. Angiosperms Examine several fruits. Draw and label the parts of one fruit.

PART B: CLEANUP AND DISPOSAL

10. Dispose of broken glass in the designated waste containers. Do not put lab materials in the trash unless your teacher tells you to do so.

11. Wash your hands thoroughly before you leave the lab and after you finish all work.

Analyze and Conclude

1. Analyzing Information How are bryophytes different from the other major groups of plants?

The gametophyte is the most noticeable stage of the moss life cycle; the

sporophyte is the most noticeable stage in other groups of plants.

_____

Surveying Plant Diversity *continued*

2. Recognizing Patterns How do the gametophytes of gymnosperms and angiosperms differ from the gametophytes of bryophytes and ferns?

Gymnosperm and angiosperm gametophytes develop within tissues of the

sporophyte; bryophyte and fern gametophytes are distinct plants.

_____

3. Drawing Conclusions What structures are present in both gymnosperms and angiosperms but absent in both bryophytes and ferns?

pollen and seeds
_____

4. Evaluating Hypotheses Dispersal is the main function of fruits in angiosperms. Defend or refute this hypothesis. List observations you made during this lab to support your position.

Answers will vary. Students should recognize that fruits often have struc-

tures that promote seed dispersal by wind, water, animals, or gravity. Many

others have fleshy, edible parts that encourage dispersal by animals.

_____

5. Inferring Conclusions Based on their characteristics, which phylum of plants appears to be the most successful? Justify your conclusion.

Answers will vary. Students should recognize that angiosperms are the most

successful phylum of plants. The presence of flowers to attract pollinators

and fruits to disperse seeds could be offered as reasons for the success of

angiosperms. The great diversity and number of angiosperms could be

offered as evidence of their success.

_____

6. Further Inquiry Write a new question about plant diversity that could be explored with another investigation.

Answers will vary.

_____

_____

Observing the Effects of Nutrients on Vegetative Reproduction

Teacher Notes

TIME REQUIRED Day 1, 45 minutes; then 10 minutes each day for 2 weeks

SKILLS ACQUIRED
Collecting Data
Designing Experiments
Experimenting
Identifying/Recognizing Patterns
Organizing and Analyzing Data
Predicting

RATING
Easy ←—— 1   2   3   4 ——→ Hard

Teacher Prep–3
Student Setup–2
Concept Level–3
Cleanup–2

SCIENTIFIC METHODS

In this lab, students will:
Make Observations
Ask Questions
Test the Hypothesis
Analyze the Results
Draw Conclusions

MATERIALS

Materials for this lab can be ordered from WARD'S. Use the Lab Materials QuickList Software on the **One-Stop Planner CD-ROM** for catalog numbers and to create a customized list of materials for this lab.

SAFETY CAUTIONS

- Be sure that students have read and understand all of the safety rules for working in the lab.

- Remind students to read the Safety section before beginning the lab. Discuss all safety symbols and caution statements with students.

- Make sure students wear safety goggles and a lab apron.

- Make sure students wash their hands thoroughly before leaving the laboratory.

TIPS AND TRICKS

Obtain duckweed from a pond or a quiet stream. You can also purchase duck-weed from a garden center or from a biological supply company. To make a 0.1 percent fertilizer solution, place 1 g of a commercial fertilizer (with an analysis of at least 23-19-17) in a large container, and add enough distilled water to make 1 L of solution. Stir until the fertilizer is dissolved. Place enough materials for a class on a supply table. Set up labeled containers for the disposal of solutions, broken glass, and duckweed.

Before students begin the lab, conduct a class discussion in which you ask the following questions:

1. What is vegetative reproduction? (the asexual process of making new individuals from nonreproductive parts of an organism)

2. How do mineral nutrients affect the growth of vegetative structures in plants? (They provide additional elements needed for making organic macromolecules and for the growth of the vegetative parts.)

Require students to present a written procedure for their experiment and a list of all safety precautions before allowing them to gather materials for Part C of the lab.

Students should see abundant vegetative reproduction in the 2-week period.

Dilute nutrient solutions with water, and pour them down the sink.

Wrap duckweed in newspaper and place into a trash can.

ANSWERS TO BEFORE YOU BEGIN

1. *vegetative reproduction*—a type of asexual reproduction that results in the growth of new individuals from nonreproductive parts, such as leaves, stems (rhizomes, tubers, bulbs, corms), and roots; *asexual reproduction*—reproduction that occurs without the union of gametes that involve a single parent producing offspring that are genetically identical to the parent; *mineral nutrients*—elements and inorganic compounds that are needed for plant growth and development

Exploration Lab

Observing the Effects of Nutrients on Vegetative Reproduction

SKILLS

- Using scientific processes
- Observing
- Graphing and analyzing data

OBJECTIVES

- **Identify** the structures of duckweed.
- **Compare** vegetative reproduction of duckweed in different nutrient solutions.

MATERIALS

- safety goggles
- lab apron
- duckweed culture
- 5 Petri dishes
- stereomicroscope or hand lens
- glass-marking pen
- beakers
- pond water
- Knop's solution
- 0.1% fertilizer solution
- distilled water

SAFETY

 CAUTION: Always wear safety goggles and a lab apron to protect your eyes and clothing.

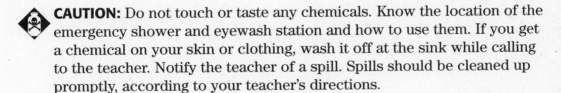 **CAUTION:** Do not touch or taste any chemicals. Know the location of the emergency shower and eyewash station and how to use them. If you get a chemical on your skin or clothing, wash it off at the sink while calling to the teacher. Notify the teacher of a spill. Spills should be cleaned up promptly, according to your teacher's directions.

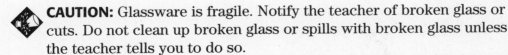 **CAUTION:** Glassware is fragile. Notify the teacher of broken glass or cuts. Do not clean up broken glass or spills with broken glass unless the teacher tells you to do so.

| Observing the Effects of Nutrients on Vegetative Reproduction *continued*

Before You Begin

Duckweed is a common aquatic plant. Like many flowering plants, duckweed reproduces readily by **vegetative reproduction,** which is a type of **asexual reproduction.** As individual plants grow, they divide into smaller individuals. Several individuals may remain joined together, forming a mat. All plants require certain **mineral nutrients,** such as nitrogen, phosphorus, and potassium, for the growth of vegetative parts. In this lab, you will investigate the effect of nutrients on the vegetative reproduction of duckweed.

1. Write a definition for each boldface term in the paragraph above. Use a separate sheet of paper. **Answers appear in the Teacher's Notes.**

2. Based on the objectives for this lab, write a question you would like to explore about vegetative reproduction in duckweed.

   For example: Which nutrients are needed for vegetative reproduction in

   duckweed?

Procedure

PART A: MAKE OBSERVATIONS

1. Place a duckweed plant in a Petri dish. Then place a few drops of water on the plant.

2. Observe the duckweed plant with a stereomicroscope or a hand lens. Sketch what you see. Label the structures that you recognize.

PART B: DESIGN AN EXPERIMENT

3. Work with members of your lab group to explore one of the questions written for step 2 of **Before You Begin.** To explore the question, design an experiment that uses the materials listed for this lab.

You Choose

As you design your experiment, decide the following:

   a. what question you will explore

   b. what your hypothesis will be

   c. what solutions to test

   d. how much of each solution to use

   e. how many individuals to use for each test

   f. what your control will be

   g. how you will judge which solution is the best

   h. what data to record in your data table

Observing the Effects of Nutrients on Vegetative Reproduction *continued*

4. Write a procedure for your experiment. Make a list of all the safety precautions you will take. Have your teacher approve your procedure and safety precautions before you begin the experiment.

PART C: CONDUCT YOUR EXPERIMENT

5. Put on safety goggles and a lab apron.

6. Set up your experiment. **CAUTION: Nutrient solutions are mild eye irritants. Avoid contact with your skin and eyes.** Complete step 8.

7. Conduct your experiment and collect data for two weeks.

PART D: CLEANUP AND DISPOSAL

8. Dispose of solutions, broken glass, and duckweed in the designated waste containers. Do not pour chemicals down the drain or put lab materials in the trash unless your teacher tells you to do so.

9. Clean up your work area and all lab equipment. Return lab equipment to its proper place. Wash your hands thoroughly before you leave the lab and after you finish all work.

Analyze and Conclude

1. **Summarizing Results** Compare the appearance of plants growing in each nutrient solution with that of the plants in distilled water. Explain your observations.

Plants grown in Knop's solution should be the largest and darkest green.

Plants grown in pond water should also look healthy and green. Plants grown

in dilute fertilizer solution or distilled water should be smaller and lighter

green.

2. **Analyzing Data** In which Petri dish did the greatest amount of growth (increase in numbers) take place?

The greatest amount of growth should occur in the dish with Knop's solution.

3. **Analyzing Results** In which Petri dish did the least amount of growth take place?

The least amount of growth should occur in distilled water.

4. **Evaluating Hypotheses** Did the results you observed agree with your hypothesis? If not, how are they different?

Answers will vary. Students should state how their results differ from their

hypotheses.

Observing the Effects of Nutrients on Vegetative Reproduction *continued*

5. Recognizing Patterns As the number of new duckweed plants in a particular group increased, what happened to the group of plants?

The groups divide as they get larger.

6. Graphing Data Make a graph of your data. Label the *y*-axis "Number of plants," and the *x*-axis "Days." Use a different color to represent each solution you tested.

Answer will vary.

7. Drawing Conclusions What factors regulate the rate of vegetative reproduction in duckweed?

Students should recognize that more mineral nutrients will produce more

growth.

8. Evaluating Methods Why are the new duckweed plants produced by vegetative reproduction genetically the same as the parent plant?

Vegetative reproduction involves mitotic cell division, and mitosis results in

the production of genetically identical cells.

9. Further Inquiry Write a new question about vegetative reproduction in duckweed that could be explored with another investigation.

Answers will vary.

Separating Plant Pigments

Teacher Notes

TIME REQUIRED 30 minutes

SKILLS ACQUIRED
Collecting Data
Constructing Models
Experimenting
Measuring

RATING
Easy ◄——— 1 —— 2 —— 3 —— 4 ——► Hard

Teacher Prep–3
Student Setup–2
Concept Level–2
Cleanup–2

SCIENTIFIC METHODS

In this lab, students will:
Make Observations
Analyze the Results
Draw Conclusions

MATERIALS

Materials for this lab can be ordered from WARD'S. Use the Lab Materials QuickList Software on the **One-Stop Planner CD-ROM** for catalog numbers and to create a customized list of materials for this lab. For every 30 students, order one WARD'S "Chromatography of Simulated Plant Pigments" kit. WARD'S refill kit contains all the consumable items. Set up labeled containers for the disposal of solutions, broken glass, and chromatograms.

SAFETY CAUTIONS

Review all safety symbols with students before beginning the lab. Make sure students wear safety goggles and a lab apron. Caution students to use care when working with the stock solutions so that they do not spill. Caution students to take care when using the capillary tubes because they are breakable. Make sure students wash their hands before leaving the lab.

TIPS AND TRICKS

Review the procedure for creating a chromatogram, on which pigments show up as colored streaks. Students'chromatograms will be almost identical to those of real plant pigments. Have students practice using a capillary tube to ensure getting a tiny drop of extract on the chromatography paper. Emphasize the importance of using a pencil when marking a chromatogram. Emphasize the importance of placing only the tip of the chromatography paper in the solvent. To prevent contamination of the solvent, the pigment should never make direct contact with the solvent. Tell students to mark the distances traveled by the pigments with a pencil as soon as the chromatogram is removed from the solvent. The simulated plant pigments fade over time. Allow students to use calculators to calculate Rf values.

ANSWERS TO BEFORE YOU BEGIN

1. *pigments*—molecules that absorb some colors of light and reflect others; *solvents*—chemicals that can dissolve other chemicals; *paper chromatography*—technique of separating a mixture of chemicals by making them pass through a strip of paper in which the chemicals move at different speeds; *Rf*—the ratio of the distance a pigment moves relative to the distance a solvent moves; *chlorophyll a*—green photosynthetic pigment found in plants; *chlorophyll b*—green photosynthetic pigment found in plants; *carotene*—orange pigment found in plants; *xanthophyll*—yellow pigment found in plants

Skills Practice Lab

Separating Plant Pigments

SKILLS
- Performing paper chromatography
- Calculating

OBJECTIVES
- **Separate** the pigments that give a leaf its color.
- **Calculate** the R_f value for each pigment.
- **Describe** how paper chromatography can be used to study plant pigments.

MATERIALS
- safety goggles
- lab apron
- strip of chromatography paper
- scissors
- metric ruler
- pencil
- capillary tube
- drop of simulated plant pigments extract
- 10 mL graduated cylinder
- 5 mL of chromatography solvent
- chromatography chamber

SAFETY

 CAUTION: Always wear safety goggles and a lab apron to protect your eyes and clothing.

 CAUTION: Do not touch or taste any chemicals. Know the location of the emergency shower and eyewash station and how to use them. If you get a chemical on your skin or clothing, wash it off at the sink while calling to the teacher. Notify the teacher of a spill. Spills should be cleaned up promptly, according to your teacher's directions.

 CAUTION: Glassware is fragile. Notify the teacher of broken glass or cuts. Do not clean up broken glass or spills with broken glass unless the teacher tells you to do so.

| Separating Plant Pigments *continued*

Before You Begin

Pigments produce colors by reflecting some colors of light and absorbing or transmitting others. Pigments can be removed from plant tissues using **solvents,** chemicals that dissolve other chemicals. The pigments can then be separated from the solvent and from each other by using **paper chromatography.** The word *chromatography* comes from the Greek words *chromat*, which means "color," and *graphon*, which means "to write." The R_f is the ratio of the distance that a pigment moves relative to the distance that a solvent moves. Since the R_f for a compound is constant, scientists can use it to identify compounds. In this lab, you will learn how to use paper chromatography to separate a mixture of pigments.

1. Write a definition for each boldface term in the previous paragraph and for each of the following terms: **chlorophyll a, chlorophyll b, carotene, xanthophyll.** Use a separate sheet of paper. **Answers appear in the Teacher's Notes.**

2. You will be using the data table provided to record your data.

3. Based on the objectives for this lab, write a question you would like to explore about plant pigments or paper chromatography.

   Answers will vary. For example: How many different pigments are found in

   the mixture?

Procedure
PART A: MAKING A CHROMATOGRAM

1. Put on safety goggles and a lab apron. Use scissors to cut the bottom end of a strip of chromatography paper to a tapered end. **CAUTION: Sharp or pointed objects may cause injury. Handle scissors carefully.**

2. Draw a faint pencil line 1 cm above the pointed end of the paper strip. Use a capillary tube to apply a tiny drop of the simulated plant pigments extract on the center of the line.

3. Pour 5 mL of chromatography solvent into a chromatography chamber. Pull the chromatography paper through the opening of the cap, and adjust the length of the strip so that a small portion of the tip end is immersed in the solvent. DO NOT immerse the pigment in the solvent.

4. Place the cap over the chromatography chamber. Carefully bend the end of the strip of chromatography paper over the cap. Be sure that the strip does not touch the walls of the chamber.

5. Remove the strip from the chromatography chamber when the solvent nears the top of the chamber (within 5–7 minutes).

Separating Plant Pigments *continued*

6. With a pencil, mark the position of the uppermost end of the solvent and the farthest distance each pigment moved. Measure the distance that the solvent and each pigment moved. Record your observations and measurements in your data table. Tape or glue your chromatogram to your lab report. Label the pigment colors.

Sample Data:

Data Table

Band no.	Color	Pigment	Migration (in mm)	R_f value
1 (top)	orange	carotene	37	0.92
2	yellow	xanthophyll	29	0.72
3	light green	chlorophyll a	17	0.42
4	dark green	chlorophyll b	8	0.20
Solvent	—	—	40	—

7. Use the formula below to calculate and record the R_f for each pigment.

$$R_f = \frac{Distance\ substance\ (pigment)\ traveled}{distance\ solvent\ traveled}$$

PART B: CLEANUP AND DISPOSAL

8. Dispose of chromatography paper, solutions, and broken glass in the designated waste containers. Do not pour chemicals down the drain or put lab materials in the trash unless your teacher tells you to do so.

9. Clean up your work area and all lab equipment. Return lab equipment to its proper place. Wash your hands thoroughly before you leave the lab and after you finish all work.

Analyze and Conclude

1. Summarizing Results Describe what happened to the simulated plant pigments during the lab.

The greenish mixture separated into four different colors: dark green, light

green, yellow, and orange. The pigments moved different distances along the

strip of chromatography paper. The orange moved the farthest distance, and

the dark green moved the shortest distance.

_____

Name _____ Class _____ Date _____

Separating Plant Pigments *continued*

2. Analyzing Data How do your R_f values compare with those of your classmates?

Answers will vary. Students should conclude that the R_f values are the same

for each pigment separated.

_____

3. Inferring Conclusions What is a chromatogram?

A chromatogram is a strip of chromatography paper that has pigments

separated from a mixture along its length.

_____

4. Further Inquiry Write a new question about plant pigments that could be explored with another investigation.

Answers will vary. For example: Can pigments in other substances such as

fruit juices and ink be separated with the same solvent?

_____

Comparing Bean and Corn Seedlings

Teacher Notes

TIME REQUIRED 30-minute lab periods

SKILLS ACQUIRED
Collecting Data
Inferring
Interpreting
Measuring

RATING

Easy ← 1   2   3   4 → Hard

Teacher Prep–2
Student Setup–2
Concept Level–2
Cleanup–1

SCIENTIFIC METHODS

In this lab, students will:
Make Observations
Ask Questions
Analyze the Results

MATERIALS

Materials for this lab can be ordered from WARD'S. Use the Lab Materials QuickList Software on the **One-Stop Planner CD-ROM** for catalog numbers and to create a customized list of materials for this lab.

SAFETY CAUTIONS

Be sure that students have read and understand all of the safety rules for working in the lab. Caution students to use scalpels carefully to avoid injuring themselves and others. Make sure students wash their hands before leaving the lab.

TIPS AND TRICKS

Bean seeds and corn kernels may be purchased at a nursery, hardware store, or feed store, or they may be ordered from a biological supply company. Soak bean seeds and corn kernels prior to this investigation. Set up labeled containers for the disposal of solutions, broken glass, and seedlings. This investigation must be done in three 30-minute periods over 5 days. The seed coat of the bean should be easy to remove if the seeds have soaked long enough.

ANSWERS TO BEFORE YOU BEGIN

1. *seed*—reproductive structure of a plant that consists of an embryo, a stored food supply, and a seed coat

   embryo—an early stage in the development of an organism

   cotyledons—structures in a seed that are part of the embryo and function in its nourishment

   root—portion of a plant that anchors the plant, absorbs water and mineral nutrients, and usually grows belowground

   monocot—plant characterized by seeds with one cotyledon

   endosperm—triploid (3n) nutritive tissue that develops from the union of a sperm with two nuclei in an ovule

   dicot—plant characterized by seeds with two cotyledons

   germinate—to begin growth, as in a seed

   seed coat—protective outer covering of a seed

   seedling—a young plant that has developed from a germinating seed

Comparing Bean and Corn Seedlings

SKILLS

- Comparing
- Drawing
- Relating

OBJECTIVES

- **Observe** the structures of bean seeds and corn kernels.
- **Compare** and **contrast** the development of bean embryos and corn embryos as they grow into seedlings.

MATERIALS

- 6 bean seeds soaked overnight
- stereomicroscope
- 6 corn kernels soaked overnight
- scalpel
- paper towels
- 2 rubber bands
- 150 mL beakers (2)
- glass-marking pen
- metric ruler

Before You Begin

A **seed** contains an inactive plant **embryo.** A plant embryo consists of one or more **cotyledons,** an embryonic shoot, and an embryonic **root.** Seeds also contain a supply of nutrients. In **monocots,** the nutrients are contained in the **endosperm.** In **dicots,** the nutrients are transferred to the cotyledons as seeds mature. A seed **germinates** when the embryo begins to grow and breaks through the protective **seed coat.** The embryo then develops into a young plant, or **seedling.** In this lab, you will examine bean seeds and corn kernels and then germinate them to observe the development of their seedlings.

1. Write a definition for each boldface term in the paragraph above. Use a separate sheet of paper. **Answers appear in the Teacher's Notes.**

2. Based on the objectives for this lab, write a question you would like to explore about seedling development.

   Answers will vary. For example: What part of a seedling emerges first?

   _____

   _____

Comparing Bean and Corn Seedlings *continued*

Procedure

PART A: OBSERVING SEED STRUCTURE

1. Remove the seed coat of a bean seed, and separate the two fleshy halves of the seed.

2. Locate the embryo on one of the halves of the seed. Examine the bean embryo with a stereomicroscope. Draw the embryo, and label the parts you can identify.

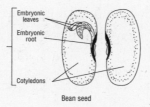

Bean seed

3. ◆ Examine a corn kernel, and locate a small light-colored oval area. **CAUTION: Sharp or pointed objects may cause injury. Handle scalpels carefully.** Use a scalpel to cut the kernel in half along the length of this area.

4. Locate the corn embryo, and examine it with a stereomicroscope. Draw the embryo, and label the parts you can identify.

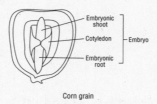

Corn grain

PART B: OBSERVING SEEDLING DEVELOPMENT

5. Fold a paper towel in half. Set five corn kernels on the paper towel. Roll up the paper towel, and put a rubber band around the roll. Stand the roll in a beaker with 1 cm of water in the bottom. Add water to the beaker as needed to keep the paper towels wet, but do not allow the corn kernels to be covered by water.

6. Repeat step 5 with five bean seeds.

7. After three days, unroll the paper towels and examine the corn and bean seedlings. Use a glass-marking pen to mark the roots and shoots of the developing seedlings. Starting at the seed, make a mark every 0.5 cm along the root of each seedling. And again starting at the seed, make a mark every 0.5 cm along the stem of each seedling.

Comparing Bean and Corn Seedlings *continued*

8. Draw a corn seedling and a bean seedling in your lab report. Label the parts of each seedling. Also show the marks you made on each seedling, and indicate the distance between the marks.

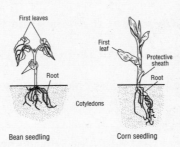

First leaves

Root

Cotyledons

Bean seedling

First leaf

Protective sheath

Root

Corn seedling

9. Using a fresh paper towel, roll up the seeds, place the rolls in the beakers, and add fresh water to the beakers.

10. After two more days reexamine the seedlings. Measure the distance between the marks. Repeat step 8.

PART C: CLEANUP AND DISPOSAL

11. Dispose of seeds, broken glass, and paper towels in the designated waste containers. Do not put lab materials in the trash unless your teacher tells you to do so.

12. Clean up your work area and all lab equipment. Return lab equipment to its proper place. Wash your hands thoroughly before you leave the lab and after you finish all work.

Analyze and Conclude

1. Relating Concepts Corn and beans are often cited as representative examples of monocots and dicots, respectively. Relate the seed structure of each to the terms *monocotyledon* and *dicotyledon*.

Corn seeds have only one cotyledon and thus are monocotyledons; bean

seeds have two cotyledons and thus are dicotyledons.

_____

_____

2. Summarizing Results What parts of a plant embryo were observed in all seedlings on the third day?

Embryonic leaves and roots are observed in all seedlings on the third day.

_____

Comparing Bean and Corn Seedlings *continued*

3. Drawing Conclusions In which part or parts of bean seedlings and corn seedlings do the seedlings grow in length? Explain.

Bean seedlings grow at the tips of their roots and stems. The distance between the marks has changed at the tips of the stems and the tips of the roots. Corn seedlings grow in length at the tips of their roots. They do not have a visible stem. The distance between the marks changed only on the roots.

4. Forming Hypotheses How are the tender young shoots of bean seedlings and corn seedlings protected as the seedlings grow through the soil?

Bean shoots have a hook in their embryonic stems. The hook pushes through the soil before the cotyledons, which enclose the embryonic shoot. Corn shoots are surrounded and protected by a sheath as they push through the soil.

5. Evaluating Viewpoints Defend the following statement: There are both similarities and differences in seed structure and seedling development in beans and corn.

In both beans and corn, the embryo consists of an embryonic shoot, an embryonic root, and cotyledons. As bean embryos and corn embryos develop, their shoots grow up, become green, and form leaves, while their roots grow down and do not become green. Bean seeds have two cotyledons; corn kernels have only one. Corn seeds have endosperm; mature bean seeds do not. The shoots of beans hook as they germinate; a corn shoot grows straight up.

6. Further Inquiry Write a new question about seedling development that could be explored with another investigation.

Answers will vary.

Skills Practice Lab

Surveying Invertebrate Diversity

Teacher Notes

TIME REQUIRED Two 45-minute periods

SKILLS ACQUIRED

Classifying
Communicating
Identifying/Recognizing Patterns
Inferring
Interpreting

RATING

Easy ← 1 2 3 4 → Hard

Teacher Prep–1
Student Setup–1
Concept Level–2
Cleanup–1

SCIENTIFIC METHODS

In this lab, students will:
Make Observations
Ask Questions
Draw Conclusions
Communicate the Results

MATERIALS

Materials for this lab can be ordered from WARD'S. Use the Lab Materials QuickList Software on the **One-Stop Planner CD-ROM** for catalog numbers and to create a customized list of materials for this lab.

SAFETY CAUTIONS

Caution students to use extreme care when observing the preserved specimens. The preservative can leak if the jars are tilted. Since some animals may be preserved in glass jars, students should be careful not to break the jars. Tell students to wash their hands if they come into contact with the preservative. Also remind students to wash their hands before leaving the laboratory.

TIPS AND TRICKS

Have students read the lab before they come to class. Before class, have students make their data table and an answer sheet for the questions at each station. This will make it easier for students to answer the questions, because not all students will start with Station 1. Have materials on the lab tables when students arrive. Students should work in groups of two to four and move clockwise from station to station.

325

ANSWERS TO BEFORE YOU BEGIN

1. *invertebrates*—animals without backbones; *body plan*—the shape, symmetry, and internal organization of the body; *radial symmetry*—symmetry in all directions about a central axis; *bilateral symmetry*—two-sided symmetry; *dorsal*—the top side; *ventral*—the bottom side; *anterior*—the front end; *posterior*—the rear end; *cephalization*—the anterior concentration of sensory structures and nerves; *segmentation*—the division of the body into repeating, similar units

2. An acoelomate has no body cavity. A pseudocoelomate has a body cavity between the mesoderm and endotherm. A coelomate has a body cavity in the mesoderm.

Name _____ Class _____ Date _____

Surveying Invertebrate Diversity

SKILLS

- Observing
- Comparing

OBJECTIVES

- **Observe** the similarities and differences among groups of invertebrates.
- **Relate** the structural adaptations of invertebrates to their evolution.

MATERIALS

- safety goggles
- lab apron
- preserved or living specimens of invertebrates
- prepared slides of sponges, hydras, planarians, and nematodes
- compound microscopes
- hand lenses or stereomicroscopes
- probes

SAFETY

  CAUTION: Always wear safety goggles and a lab apron to protect your eyes and clothing.

 CAUTION: Do not touch or taste any chemicals. Know the location of the emergency shower and eyewash station and how to use them. If you get a chemical on your skin or clothing, wash it off at the sink while calling to the teacher. Notify the teacher of a spill. Spills should be cleaned up promptly, according to your teacher's directions.

CAUTION: Glassware is fragile. Notify the teacher of broken glass or cuts. Do not clean up broken glass or spills with broken glass unless the teacher tells you to do so.

Before You Begin

Invertebrates include all animals except those with backbones. Every phylum of the kingdom Animalia except the phylum Chordata consists only of invertebrates. In this lab, you will examine representatives of eight phyla of animals. You will see many similarities and differences in **body plan**—shape, symmetry, and internal organization.

1. Write a definition for each boldface term in the paragraph above and for the following terms: radial symmetry, bilateral symmetry, dorsal, ventral, anterior, posterior, cephalization, segmentation. Use a separate sheet of paper.
   Answers appear in the Teacher's Notes.

2. Describe the three basic body plans found in animals. Use a separate sheet of paper. **Answers appear in the Teacher's Notes.**

3. You will be using the data table provided to record your data.

4. Based on the objectives for this lab, write a question you would like to explore about the characteristics of invertebrates.

Students' questions will vary. For example: What characteristics are shared

by all invertebrates?

Procedure

PART A: CONDUCTING A SURVEY

1. Put on safety goggles and a lab apron.

2. Visit each invertebrate station, and examine the specimens there. Answer the questions, and record observations in your data table.

Data Table				
Phylum	**Symmetry**	**Body plan**	**Other**	**Examples**

3. Sponges Examine each specimen.

    a. Describe the shape of a sponge.

    Sponges have an irregular shape; they are asymmetrical.

    b. What do you think is the role of the many holes, or pores, in a sponge?

    The pores allow water, which carries in food particles and oxygen and

    carries out wastes, to flow through the sponge.

    c. Examine a prepared slide of a sponge with a compound microscope. What do you notice about the organization of the cells in sponges?

    The cells in a sponge are of several types that are loosely arranged in

    several layers but are not organized into tissues.

Surveying Invertebrate Diversity *continued*

4. **Cnidarians** Examine each specimen.

   a. Divide the cnidarian specimens into two groups. What feature did you use to make your division?

   body shape; There are polyps and medusas.

   b. How many body openings does a cnidarian have?

   one

   c. Examine a prepared slide of a hydra. What do you notice about the organization of the cells in cnidarians?

   The cells are of several types and are in two tissue layers.

5. **Flatworms and Roundworms** Examine each specimen.

   a. How does a flatworm differ from a roundworm in external appearance?

   Flatworms are flat and vary in length. Roundworms have long, tubular

   bodies.

   b. Do any of the worms appear to be segmented? Explain.

   Students may respond "yes" if they view a tapeworm because its body

   looks somewhat segmented. Explain that only coelomates show true seg-

   mentation. Tapeworms do have a series of similar sections, but this is

   different from true segmentation.

   c. Examine prepared slides of planarians and nematodes. How many body openings does each have?

   Planarians have one; nematodes have two.

6. **Mollusks** Examine each specimen.

   a. In what ways do the mollusks differ in external appearance?

   Mollusks may have a two-part shell, single shell, or no shell.

   b. Which group of mollusks has the most noticeable "feet"?

   Gastropods—the foot is a large muscular structure.

Surveying Invertebrate Diversity *continued*

7. **Annelids** Examine each specimen.

  a. How are an earthworm and a leech similar? How are they different?

  Both have bilateral symmetry and a body divided into many segments.

  Earthworms are tubular and much longer than they are wide; leeches are

  flattened and not nearly as long.

  b. Describe any differences you see in the segments of the annelid worm.

  Segments at the anterior and posterior ends are smaller than those in the

  middle.

8. **Arthropods** Examine each specimen.

  a. What characteristic do you observe in all arthropod appendages?

  They are jointed.

  b. How does the number of walking legs differ among these arthropods?

  Millipedes have two pairs of legs per body segment, centipedes have one

  pair of legs per body segment, insects have a total of three pairs, spiders

  have four, and decapod crustaceans have five.

9. **Echinoderms** Examine each specimen.

  a. The word *echinoderm* means "spiny skin." Why is this name appropriate?

  Most echinoderms are covered by numerous spines that grow out of the

  skin.

  b. What does an echinoderm's ventral surface look like?

  The ventral surface has several grooves in which there are numerous tube

  feet.

PART B: CLEANUP AND DISPOSAL

10. Dispose of broken glass in the designated waste containers. Do not put lab materials in the trash unless your teacher tells you to do so.

11. Wash your hands thoroughly before you leave the lab and after you finish all work.

Surveying Invertebrate Diversity *continued*

Analyze and Conclude

1. Summarizing Data Which animal phyla show cephalization, and which do not?

All but the sponges, cnidarians, and echinoderms exhibit cephalization.

2. Recognizing Patterns What type of symmetry is found with cephalization?

bilateral symmetry

3. Recognizing Patterns What characteristics do annelids and arthropods share?

segmentation, bilateral symmetry, and cephalization

4. Analyzing Methods Were you able to identify the type of body plan found in all of the specimens? Explain.

No, body plans cannot be determined by examining an animal's exterior.

5. Further Inquiry Write a new question about invertebrates that could be explored with another investigation.

Answers will vary. For example: How does an arthropod's outer covering

differ from a worm's?

Observing Hydra Behavior

Teacher Notes

TIME REQUIRED 45 minutes

SKILLS ACQUIRED
Collecting Data
Communicating
Designing Experiments

RATING
Easy ◄――1――2――3――4――► Hard

Teacher Prep–2
Student Setup–2
Concept Level–3
Cleanup–2

SCIENTIFIC METHODS

In this lab, students will:
Make Observations
Ask Questions
Analyze the Results
Draw Conclusions

MATERIALS

Materials for this lab can be ordered from WARD'S. Use the Lab Materials QuickList Software on the **One-Stop Planner CD-ROM** for catalog numbers and to create a customized list of materials for this lab.

SAFETY CAUTIONS

Caution students to handle glassware carefully to avoid injuring themselves or others.

TIPS AND TRICKS

Demonstrate how to pick up hydras and *Daphnia* with a medicine dropper. If the tip of the dropper is too small, the dropper can be inverted. Instead of *Daphnia*, you may also use brine shrimp or a thread soaked in anything that has glutathione (e.g., liver) to elicit capture behavior in the hydra. Discuss some of the basic principles of sensory cells, neurons, and effector cells, relating the information to the way hydras capture prey. To observe whether a hydra responds to a chemical stimulus (nutrient), the students can hold a pennant-shaped piece of filter paper with forceps and move the long tip of the pennant near, but not touching, the hydra's tentacles (as a control). After observing and recording the hydra's response to the filter paper, they can dip the same piece of filter paper in beef broth and repeat the procedure. To observe how a hydra responds to touch, students can use the long tip of a clean pennant-shaped piece of filter paper and touch the hydra's tentacles, mouth, disk, and stalk. To observe feeding behavior, students can use a medicine dropper to transfer live *Daphnia* to the well with the hydra on the microscope slide.

ANSWERS TO BEFORE YOU BEGIN

1. *cnidarian*—an organism of the phylum Cnidaria that has a hollow gut with a single opening and flexible, fingerlike tentacles

   hydra—common freshwater cnidarian with a body stalk and tentacles surrounding the mouth

   behavior—how an animal responds to a stimulus

   nematocysts—cells found on a cnidarian that are used in defense and in capturing prey

Exploration Lab

Observing Hydra Behavior

SKILLS

- Using scientific processes
- Observing

OBJECTIVES

- **Observe** a hydra finding and capturing prey.
- **Determine** how a hydra responds to stimuli.

MATERIALS

- silicone culture gum
- microscope slide
- 2 medicine droppers
- *Hydra* culture
- *Daphnia* culture
- concentrated beef broth
- filter paper cut into pennant shapes
- forceps
- stereomicroscope

Before You Begin

Cnidarians are carnivorous animals. A common cnidarian is **Hydra**, a freshwater organism that feeds on smaller freshwater animals, such as water fleas (*Daphnia*). Hydras find food by responding to stimuli, such as chemicals and touch. The way an animal responds to stimuli is called **behavior.** The tentacles of a cnidarian are armed with **nematocysts,** which are used in defense and in capturing prey. When a hydra receives stimuli from potential prey, its nematocysts spring out and harpoon or entangle the prey. In this lab, you will observe the feeding behavior of hydras to determine how they find and capture prey.

1. Write a definition for each boldface term in the paragraph above. Use a separate sheet of paper. **Answers appear in the Teacher's Notes.**

2. Based on the objectives for this lab, write a question you would like to explore about the feeding behavior of hydras.

   Answers will vary. For example: How does a hydra respond to a *Daphnia* that

   swims close to it?

Name _____ Class _____ Date _____

Observing Hydra Behavior *continued*

Procedure

PART A: MAKE OBSERVATIONS

1. To make an experimental pond for observing hydras, squeeze out a long piece of silicone culture gum. Arrange it to form a circular well on a microscope slide. **CAUTION: Glassware is fragile. Notify the teacher promptly of any broken glass or cuts.**

2. With a medicine dropper, gently transfer a hydra from its culture dish to the well on the slide, making sure the water covers the animal. **CAUTION: Handle hydras gently to avoid injuring them.** Allow the hydra to settle, then examine it under the high power of a stereomicroscope. Draw a hydra and label the body stalk, mouth, and tentacles.

PART B: DESIGN AN EXPERIMENT

3. Work with the members of your lab group to explore one of the questions written for step 2 of **Before You Begin.** To explore the question, design an experiment that uses the materials listed for this lab.

4. Write a procedure for your experiment. Make a list of all the safety precautions you will take. Have your teacher approve your procedure and safety precautions before you begin the experiment.

You Choose

As you design your experiment, decide the following:

   a. what question you will explore

   b. what hypothesis you will test

   c. how to observe a hydra's feeding behavior

   d. how to test a hydra's response to a stimulus, such as a chemical or a touch

   e. what your test groups and controls will be

   f. what to record in your data table

PART C: CONDUCT YOUR EXPERIMENT

5. Set up and carry out your experiment. **CAUTION: Handle hydras gently to avoid injuring them.**

6. Allow hydras to settle before exposing them to a test condition. If your hydra does not respond after a few minutes, obtain another hydra from the culture dish. Repeat your procedure.

Name _____ Class _____ Date _____

Observing Hydra Behavior *continued*

PART D: CLEANUP AND DISPOSAL

7. Dispose of lab materials and broken glass in the designated waste containers. Put hydras and daphnias in the designated containers. Do not put lab materials in the trash unless your teacher tells you to do so.

8. Clean up your work area and all lab equipment. Return lab equipment to its proper place. Wash your hands thoroughly before you leave the lab and after you finish all work.

Analyze and Conclude

1. **Analyzing Results** Describe a hydra's response to chemicals (beef broth).

   The hydra should show a feeding response to the beef broth, which includes

   expansion of its mouth, movement of its tentacles, elongation of its body, and

   the release of nematocysts.

2. **Analyzing Results** Describe a hydra's response to touch.

   The hydra contracts its body.

3. **Drawing Conclusions** How does a hydra detect its prey?

   A hydra responds to chemicals in the water to detect the presence of prey.

4. **Justifying Conclusions** Give evidence to support your conclusion about how hydras detect prey.

   The hydra exhibited predatory behavior in response to the beef broth. Yet the

   hydra displayed nonpredatory behavior when touched.

5. **Inferring Conclusions** Based on your observations, how do you think a hydra behaves when it detects a threat in its natural habitat?

   Based on the hydra's response to touch, students should conclude that a hydra

   contracts its body when threatened.

6. **Inferring Conclusions** What happens to food that has not been digested by a hydra?

   Undigested food is released from the hydra's mouth.

Observing Hydra Behavior *continued*

7. Inferring Conclusions How is a hydra adapted to a sedentary lifestyle?

Tentacles and nematocysts enable a hydra to capture prey that drift past it in

slow-moving water. Its ability to contract when touched protects the hydra from

predators.

_____

8. Further Inquiry Write a new question about the behavior of hydras that could be explored with another investigation.

Answers will vary. For example: How does a hydra respond to light?

_____

Observing Characteristics of Clams

Teacher Notes

TIME REQUIRED 45 minutes

SKILLS ACQUIRED
Inferring
Interpreting
Measuring
Predicting

RATING

Easy ◄——1——2——3——4——► Hard

Teacher Prep–2
Student Setup–2
Concept Level–2
Cleanup–2

SCIENTIFIC METHODS

In this lab, students will:
Make Observations
Ask Questions
Draw Conclusions

MATERIALS

Materials for this lab can be ordered from WARD'S. Use the Lab Materials QuickList Software on the **One-Stop Planner CD-ROM** for catalog numbers and to create a customized list of materials for this lab.

SAFETY CAUTIONS

Make sure students wear safety goggles, lab aprons, and disposable gloves throughout the investigation.

Before students begin step 8, remind them to keep hydrochloric acid (HCl) away from skin and clothing. If acid is spilled, thoroughly flush all affected areas with running water and mop up the spill with cloths designed for chemical spill cleanup. Avoid having students handle acid spills.

TIPS AND TRICKS

Keep the clams in a freshwater aquarium with green algae and a layer of sand until lab time. Review the mollusk body plan before the lab. Before students begin step 7, tell them to use extreme care when chipping clam shells with their scalpel. Have students use a scalpel that has a permanent blade. Broken glass and pieces of clam shell can be thrown in the trash. Unused 0.1 M HCl solution can be neutralized by carefully adding 0.1 M NaOH until the pH is between 6 and 8. The resulting solution can then be poured down the drain.

ANSWERS TO BEFORE YOU BEGIN

1. *mollusks*—animals with a soft body that secrete a shell containing calcium carbonate; *foot*—part of the mollusk body that extends from the shell and aids in locomotion; *incurrent siphon*—tube that takes water into a bivalve's shell; *excurrent siphon*—tube that expels water from a bivalve's shell; *calcium carbonate*—mineral found in a mollusk's shell; *mantle*—tissue that lines a mollusk's shell and secretes the shell; *umbo*—the oldest part of a bivalve's shell

Skills Practice Lab

Observing Characteristics of Clams

SKILLS

• Observing

• Testing for the presence of a chemical

OBJECTIVES

• **Observe** the behavior of a live clam.

• **Examine** the structure and composition of a clam shell.

MATERIALS

• safety goggles

• lab apron

• live clam

• small beaker or dish

• eyedropper

• food coloring

• glass stirring rod

• clam shell

• Petri dish

• scalpel

• stereomicroscope

• 0.1 M HCl

SAFETY

  CAUTION: Always wear safety goggles and a lab apron to protect your eyes and clothing.

 CAUTION: Do not touch or taste any chemicals. Know the location of the emergency shower and eyewash station and how to use them. If you get a chemical on your skin or clothing, wash it off at the sink while calling to the teacher. Notify the teacher of a spill. Spills should be cleaned up promptly, according to your teacher's directions.

 CAUTION: Glassware is fragile. Notify the teacher of broken glass or cuts. Do not clean up broken glass or spills with broken glass unless the teacher tells you to do so.

Before You Begin

Clams are **mollusks,** and they have a two-part shell. The body of a clam consists of a visceral mass and a muscular **foot.** There is no definite head. Two tubes, an **incurrent siphon** and an **excurrent siphon,** extend from the body on the side opposite the foot. Like all mollusks, clams have a shell composed of **calcium carbonate.** A membrane called the **mantle** lines the shell and forms successive rings of shell as a clam grows. The **umbo** is the oldest part of a clam shell. In this lab, you will examine live clams and clam shells.

1. Write a definition for each boldface term in the paragraph. Use a separate sheet of paper. **Answers appear in the Teacher's Notes.**

2. Based on your objectives, write a question you would like to explore about clams.

   Answers will vary. For example: How does a clam respond to touch?

   _____

Procedure

PART A: OBSERVE A LIVE CLAM

1. Put on safety goggles and a lab apron.

2. Place a live clam in a small beaker or shallow dish of water. Using an eyedropper, apply two drops of food coloring near the clam.

3. Observe and record what happens to the food coloring.

   The food coloring will be taken in by the incurrent siphon. Students will

   observe colored water being expelled near the excurrent siphon.

4. Using a stirring rod, touch the clam's mantle. **CAUTION: Touch the clam gently to avoid injuring it.**

5. Observe and record the clam's response to touch.

   Answers will vary. Students will probably observe that the clam draws in its

   siphons and foot and closes its shell tightly when touched.

Name _____ Class _____ Date _____

Observing Characteristics of Clams *continued*

PART B: OBSERVE A CLAM SHELL

6. Examine the concentric growth rings on the shell. Locate the knob-shaped umbo on the shell. Count and record the number of growth rings on the clam shell.

<u>**Answers will vary. The number of growth rings will depend on the clam's age.**</u>

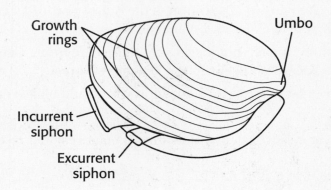

7. Place the clam shell in a Petri dish. Use a scalpel to chip away part of the shell to expose its three layers. **CAUTION: Sharp or pointed objects may cause injury. Handle scalpels carefully.** View the shell's layers with a stereomicroscope. The outermost layer protects the clam from acids in the water. The innermost layer is mother-of-pearl, the material that forms pearls.

8. The middle layer of the shell contains crystals of calcium carbonate. To test for the presence of this compound, place one drop of 0.1 M HCl on the middle layer of the shell. **CAUTION: Hydrochloric acid is corrosive. Avoid contact with skin, eyes, and clothing. Avoid breathing vapors.** If calcium carbonate is present, bubbles of carbon dioxide will form in the drop. Record your observations.

<u>**Students should observe bubble formation, indicating the presence of**</u>

<u>**calcium carbonate in the middle layer of the shell.**</u>

PART C: CLEANUP AND DISPOSAL

9. Dispose of solutions, broken glass, and pieces of clam shell in the waste containers designated by your teacher. Do not pour chemicals down the drain or put lab materials in the trash unless your teacher tells you to do so.

10. Clean up your work area and all lab equipment. Return live clams to the stock container. Return lab equipment to its proper place. Wash your hands thoroughly before you leave the lab and after you finish all work.

Observing Characteristics of Clams *continued*

Analyze and Conclude

1. Analyzing Results Find the incurrent and excurrent siphons of the clam in the illustration on the previous page. Using this information, explain your observations in step 3.

The food coloring was drawn into the clam through the incurrent siphon and

released through the excurrent siphon.

2. Drawing Conclusions What is the purpose of a clam's shell?

The shell protects the clam from injury and predators.

3. Making Predictions Based on your observations, how do you think clams respond when they are touched or threatened in their natural habitat?

Clams draw in their siphons and foot and close their shells tightly when

threatened or touched.

4. Forming a Hypothesis What does a clam take in from water that passes through its body?

Food and oxygen are taken in from the water.

5. Inferring Relationships Water that enters a clam's incurrent siphon passes over the clam's gills. How does this help the clam respire?

A mollusk's gills can extract 50 percent or more of the dissolved oxygen

from water that passes over them.

6. Further Inquiry Write a new question about clams that could be explored with another investigation.

Answers will vary. For example: What types of food do clams prefer?

Observing Pill Bug Behavior

Teacher Notes

TIME REQUIRED 40–50 minutes

SKILLS ACQUIRED
Designing Experiments
Experimenting
Collecting Data
Organizing and Analyzing Data

RATING

Easy ←——1——2——3——4——→ Hard

Teacher Prep–2
Student Setup–2
Concept Level–3
Cleanup–2

SCIENTIFIC METHODS

In this lab, students will:
Make Observations
Test their Hypotheses
Analyze the Results
Draw Conclusions

MATERIALS

Materials for this lab can be ordered from WARD'S. Use the Lab Materials QuickList Software on the **One-Stop Planner CD-ROM** for catalog numbers and to create a customized list of materials for this lab.

SAFETY CAUTIONS

Discuss all safety symbols and caution statements with students. Remind students to be careful when using scissors.

TIPS AND TRICKS

Encourage students to read the lab and complete the Before You Begin section before they come to class on the day of the lab. Remind students that although pill bugs have an exoskeleton, they are delicate creatures and should be probed very gently. Before students begin the lab, conduct a class discussion in which you ask the following questions: What are the characteristics of a crustacean? What is the ideal habitat for land-dwelling isopods?

Distribute the pill bugs to students in small paper cups. Students should perform their experiments using at least three pill bugs.

ANSWERS TO BEFORE YOU BEGIN

1. *pill bugs*—small crustaceans in the order Isopoda that live in moist terrestrial environments

crustaceans—arthropods with mandibles, two pairs of antennae, and branched appendages.

stimulus—any action or agent that causes or changes an activity in an organism

ANSWERS TO ANALYZE AND CONCLUDE

3. Sample Graph:

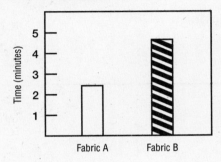

Observing Pill Bug Behavior

SKILLS

- Using scientific methods
- Observing

OBJECTIVES

- **Identify** arthropod characteristics in a pill bug.
- **Observe** the behavior of pill bugs on surfaces with different textures.
- **Infer** the adaptive advantages of pill bug behaviors.

MATERIALS

- 4 adult pill bugs
- 2 Petri dishes
- stereomicroscope or hand lens
- blunt probe
- fabrics with different textures
- scissors
- transparent tape
- clock or watch with second hand

SAFETY

  CAUTION: Always wear safety goggles and a lab apron to protect your eyes and clothing.

 CAUTION: Do not touch or taste any chemicals. Know the location of the emergency shower and eyewash station and how to use them. If you get a chemical on your skin or clothing, wash it off at the sink while calling to the teacher. Notify the teacher of a spill. Spills should be cleaned up promptly, according to your teacher's directions.

 CAUTION: Glassware is fragile. Notify the teacher of broken glass or cuts. Do not clean up broken glass or spills with broken glass unless the teacher tells you to do so.

Before You Begin

Pill bugs live in moist terrestrial environments, such as under rocks and logs. Like other **crustaceans,** pill bugs respire with gills and have hard outer shells and jointed appendages. They respond to a **stimulus,** such as light, moisture, or touch, by moving toward or away from the stimulus or by curling into a ball. In this lab, you will look for arthropod characteristics in pill bugs and observe the behavior of pill bugs on surfaces with different textures.

Name _____ Class _____ Date _____

Observing Pill Bug Behavior *continued*

1. Write a definition for each boldface term in the preceding paragraph. Use a separate sheet of paper. **Answers appear in the Teacher's Notes.**

2. Based on the objectives for this lab, write a question you would like to explore about pill bug characteristics and behavior.

   <u>**Answers will vary. For example: Do pill bugs prefer rough or smooth surfaces?**</u>

   _____

Procedure
PART A: MAKE OBSERVATIONS

1. Place a pill bug in a Petri dish, and observe it with a stereomicroscope or hand lens. Observe it from a dorsal viewpoint as well as from the side. List the characteristics that tell you the pill bug is an arthropod.

   Arthropod characteristics of pill bugs include an exoskeleton, jointed

   appendages, and a segmented body.

   _____

2. Touch the pill bug with a blunt probe. **CAUTION: Touch pill bugs gently to avoid injuring them.** Record your observations.

   When students touch a pill bug with a blunt probe, the pill bug should curl into

   a ball.

   _____

PART B: DESIGN AN EXPERIMENT

3. Work with the members of your lab group to explore one of the questions written for step 2 of **Before You Begin.** To explore the question, design an experiment that uses the materials listed for this lab.

   > **You Choose**
   >
   > As you design your experiment, decide the following:
   >
   > **a.** what question you will explore
   >
   > **b.** what hypothesis you will test
   >
   > **c.** which four different fabrics to use
   >
   > **d.** how many times to test each fabric
   >
   > **e.** the length of each test
   >
   > **f.** what your control will be
   >
   > **g.** what data to record in your data table

Observing Pill Bug Behavior *continued*

4. Write a procedure for your experiment. Make a list of all the safety precautions you will take. Have your teacher approve your procedure and safety precautions before you begin the experiment.

5. Set up and carry out your experiment. **CAUTION: Sharp or pointed objects can cause injury. Handle scissors carefully.**

PART C: CLEANUP AND DISPOSAL

6. Dispose of fabric scraps and broken glass in the designated waste containers. Put pill bugs in the designated container. Do not put lab materials in the trash unless your teacher tells you to do so.

7. Clean up your work area and all lab equipment. Return lab equipment to its proper place. Wash your hands thoroughly before you leave the lab and after you finish all work.

Analyze and Conclude

1. **Analyzing Methods** Why did you test several pill bugs in this investigation instead of just one pill bug?

   Several pill bugs were tested to improve the reliability of the data.

2. **Analyzing Results** Did all of your pill bugs show a similar pattern of movement? Explain.

   Answers may vary. The pill bugs should show a similar pattern of movement.

3. **Graphing Results** Make a graph of your data on a sheet of graph paper. Plot the average time spent on the material on the *y*-axis and the type of material on the *x*-axis. **A sample graph appears in the Teacher's Notes.**

4. **Analyzing Results** Rank the fabrics according to the total amount of time spent on them by the pill bugs.

   Answers will depend on the types of fabrics used. The highest ranking should

   go to the fabric on which the pill bugs spent the most time.

5. **Drawing Conclusions** Which fabric texture do pill bugs seem to prefer?

   Answers will depend on the types of fabrics used. Pill bugs should spend more

   time on rough-textured fabrics, such as wool, than on smooth-textured

   fabrics, such as silk.

Observing Pill Bug Behavior *continued*

6. Inferring Conclusions How is a pill bug's response to disturbances an advantage?

The pill bug's response to disturbances allows it to protect its gills and under-

side by rolling into a ball.

_____

7. Inferring Conclusions How is being able to detect surface texture helpful to pill bugs in their natural habitat?

Being able to distinguish between rough textures and smooth textures may

help pill bugs find soft substrates and avoid crawling over surfaces that

might expose them to predators.

8. Further Inquiry Write a new question about pill bugs that could be explored with another investigation.

Answers will vary. For example: Do pill bugs prefer moist or dry environments?

Do pill bugs prefer light or dark environments?

_____

Analyzing Sea Star Anatomy

Teacher Notes

TIME REQUIRED 50 minutes

SKILLS ACQUIRED

Identifying/Recognizing Patterns
Collecting Data
Interpreting
Communicating

RATING

Easy ◄——— 1 ——— 2 ——— 3 ——— 4 ———► Hard

Teacher Prep–3
Student Setup–2
Concept Level–2
Cleanup–3

SCIENTIFIC METHODS

In this lab, students will:
Make Observations
Analyze the Results
Draw Conclusions
Communicate the Results

MATERIALS

Materials for this lab can be ordered from WARD'S. Use the Lab Materials QuickList Software on the **One-Stop Planner CD-ROM** for catalog numbers and to create a customized list of materials for this lab. You will need one preserved sea star, dissection tray, set of dissecting instruments, and hand lens or dissecting microscope for each student or lab group. Set up labeled containers for the disposal of sea star body parts, dissecting pins, and gloves.

SAFETY CAUTIONS

Review all safety symbols with students before beginning the lab. Make sure students wear safety goggles, gloves and a lab apron, and wear this protective gear yourself when handling preserved specimens. Caution students to use care when working with sharp instruments. Caution students to keep their hands away from their faces during the lab. Make sure students wash their hands before leaving the lab.

TIPS AND TRICKS

Leave sea stars in a flowing water bath for at least 4 hours to remove preservative before the lab begins. Demonstrate for students safe techniques for using dissection instruments and dissecting pins. Encourage students to observe other students' specimens and to note external and internal differences between individuals of a single species.

ANSWERS TO BEFORE YOU BEGIN

1. *echinoderm*—a radially symmetrical marine invertebrate that has an endoskeleton, such as a sea star, a sea urchin, or a sea cucumber

   endoskeleton—an internal skeleton made of bone and cartilage

   five-part radial symmetry—a body plan in which five similar parts of an animal's body are organized in a circle around a central axis

   water-vascular system—a system of canals filled with a watery fluid that circulates throughout the body of an echinoderm

   coelom—a body cavity that contains the internal organs

Skills Practice Lab

Analyzing Sea Star Anatomy

SKILLS

- Observing
- Collecting data
- Inferring

OBJECTIVES

- **Observe** anatomical structures of an echinoderm.
- **Infer** function of body parts from structure.

MATERIALS

- disposable gloves
- preserved sea star
- dissection tray
- dissection scissors
- hand lens
- dissecting microscope
- forceps
- blunt probe
- sharp probe
- dissection pins

SAFETY

 CAUTION: Always wear safety goggles and a lab apron to protect your eyes and clothing.

 CAUTION: Do not touch or taste any chemicals. Know the location of the emergency shower and eyewash station and how to use them. If you get a chemical on your skin or clothing, wash it off at the sink while calling to the teacher. Notify the teacher of a spill. Spills should be cleaned up promptly, according to your teacher's directions.

 CAUTION: Glassware is fragile. Notify the teacher of broken glass or cuts. Do not clean up broken glass or spills with broken glass unless the teacher tells you to do so.

Before You Begin

Sea stars are members of the phylum Echinodermata, a group of invertebrates that also includes sand dollars, sea urchins, and sea cucumbers. **Echinoderms** share four main characteristics: an **endoskeleton, five-part radial symmetry,** a **water-vascular system,** and circulation and respiration through their **coelom.**

1. Write a definition for each boldface term above. Use a separate sheet of paper. **Answers appear in the Teacher's Notes.**

2. You will be using the data table provided to record your data.

Analyzing Sea Star Anatomy *continued*

3. Based on the objectives for this lab, write a question you would like to explore about sea star anatomy.

Answers will vary. For example: Which of the sea star's body structures are

modified for feeding?

Procedure
PART A: EXTERNAL ANATOMY

1. ◇ ◇ ◇ **CAUTION: Put on safety goggles, a lab apron, and protective gloves.** As you observe the sea star body structures, record your observations and your inference of each structure's function in the table. On a separate sheet of paper or in your lab notebook, draw and label the sea star and the structures that you observe.

Data Table
Function of Sea Star Structures

Structure	Observations	Inferred function
Madreporite		
Spine		
Skin gill		

2. Using forceps, hold a preserved sea star under running water to gently but thoroughly remove excess preservative. Then place the sea star in a dissecting tray.

3. Refer to a diagram of a sea star in your textbook to locate the madreporite on the upper surface of the sea star.

4. Use a hand lens to observe the sea star's spines. Are they distributed in any recognizable pattern? Are they exposed or covered by tissue? Are they movable or fixed?

Spines are short, scattered over the surface, exposed, and fixed.

5. Use the dissecting microscope to look for small skin gills, If any are present, describe their location and structure.

Skin gills are distributed between the spines but may not be visible

in preserved specimens.

6. Examine the sea star's lower surface. Find the mouth, and use forceps or a probe to gently move aside any soft tissues. What structures are found around the mouth?

Five pairs of movable spines surround the mouth.

7. Locate the tube feet. Describe their distribution. Using a dissecting microscope, observe and then draw a single tube foot on a separate sheet of paper.

Tube feet are arranged in two rows along the length of each arm's oral surface.

PART B: INTERNAL ANATOMY

8. **CAUTION: Scissors, probes, and pins are sharp. Use care not to puncture your gloves or injure yourself or others.** Using scissors and forceps, carefully cut the body wall away from the upper surface of one of the sea star's arms. Start near the end of the arm and work toward the center.

9. Find the digestive glands in the arm you have opened. Then, locate the short branched tube that connects the digestive glands to the pyloric stomach.

10. Cut the tube that connects the digestive glands to the stomach, and move the digestive glands out of the arm. Look for the reproductive organs.

11. Locate the two rows of ampullae that run the length of the arm.

12. Carefully remove the body wall from the upper surface of the central region of the sea star. Locate the pyloric stomach and the cardiac stomach.

13. Remove the stomachs and find the ring canal and the radial canals. In which direction does water move through these canals?

Water moves from the ring canal to the radial canals.

14. Dispose of sea stars and sea star body parts in the waste container designated by your teacher. Do not put lab materials in the trash unless your teacher tells you to do so.

15. Clean up your work area and all lab equipment. Return lab equipment to its proper place. Wash your hands thoroughly before you leave the lab and after you finish all work.

Analyze and Conclude

1. Analyzing Results What type of symmetry is found in the sea star?

Sea stars have five-part radial symmetry.

2. Inferring Relationships What is the relationship between the ampullae and the tube feet?

The ampullae are hollow and are connected to the tube feet. The ampullae

pump water into the tube feet, causing the tube feet to expand outward.

3. Making Predictions How does a sea star use its stomach during feeding?

Part of the stomach is thrust through the mouth, and digestive juices liquefy

the prey, which then is ingested.

4. Making Predictions If the ring canals and radial canals did not function properly, how would this affect the sea star's ability to move and feed?

Because the sea star's water-vascular system is essential to locomotion and

feeding, the sea star would not be able to move about or feed normally if the

ring canals and radial canals malfunctioned.

5. Further Inquiry Write a new question about echinoderms that could be explored with another investigation.

Answers will vary. For example: How do the bodies of sea urchins differ from

those of sea stars?

Comparing Hominid Skulls

Teacher Notes

TIME REQUIRED One 40-minute class period

SKILLS ACQUIRED

Collecting Data
Identifying/Recognizing Patterns
Inferring
Interpreting
Organizing and Analyzing Data

RATING

Easy ←——1——2——3——4——→ Hard

Teacher Prep–2
Student Setup–1
Concept Level–2
Cleanup–1

SCIENTIFIC METHODS

In this lab, students will:

Make Observations   Analyze the Results
Test the Hypothesis   Draw Conclusions

MATERIALS

Materials for this lab can be ordered from WARD'S. Use the Lab Materials QuickList Software on the **One-Stop Planner CD-ROM** for catalog numbers and to create a customized list of materials for this lab.

TIPS AND TRICKS

If life-size casts of skulls are available, substitute these for the diagrams provided, or use them to supplement the investigation.

The checklist of features has incorporated conversion factors of 40 for calculating square area and 1,000 for calculating cubic volume of life-size skulls. If life-size casts of skulls are available for students to measure, eliminate the factor of 40 in calculating square area. Substitute 4.2, or $4/3 \times \pi$, for the factor of 1,000 in calculating cubic volume.

ANSWERS TO BEFORE YOU BEGIN

1. *apes*—tailless primates with long arms, a broad chest, and a larger, more developed brain than monkeys; *common ancestor*—one species from which two or more species evolved; *hominids*—members of the family Hominidae of the order Primates, and are characterized by opposable thumbs, no tail, longer lower limbs, and erect bipedalism; *anatomy*—the body structure of an organism.

Exploration Lab

Comparing Hominid Skulls

SKILLS

- Measuring
- Comparing anatomical features

OBJECTIVES

- **Identify** differences and similarities between the skulls of apes and the skulls of humans.
- **Identify** differences and similarities between the fossilized skulls of hominids.
- **Classify** the features of hominid skulls as apelike, humanlike, or intermediate.

MATERIALS

- metric ruler
- protractor

Before You Begin

Modern **apes** and humans share a **common ancestor.** Much of our understanding of human evolution is based on the study of the fossilized remains of **hominids.** By studying fossilized bones and identifying similar and dissimilar structures, scientists can infer the **anatomy,** or body structure, of a species. In this lab, you will identify differences and similarities between the skulls of apes, early hominids, and humans.

1. Write a definition for each boldface term in the paragraph above. Use a separate sheet of paper. **Answers appear in the Teacher's Notes.**

2. You will be using the data table provided to record your data.

3. Based on the objectives for this lab, write a question you would like to explore about human evolution.

    Answers will vary. For example: How have facial features changed as

    hominids have evolved?

    _____

How to Interpret the Features of a Skull

Cranial capacity: Use the circles drawn on the skulls to estimate brain volume, or cranial capacity. Measure the radius of each circle in centimeters. Then cube this number, and multiply the result by 1,000 to calculate the approximate life-size cranial capacity in cubic centimeters.

Lower face area: Measure A to B and C to D in centimeters for each skull. Multiply these two numbers together, and multiply the product by 40 to approximate the life-size lower face area in square centimeters.

Brain area: Measure E to F and G to H in centimeters for each skull. Multiply these two numbers and multiply the product by 40 to approximate the life-size brain area in square centimeters.

Jaw angle: Note the two lines that come together near the nose of each skull. Use a protractor to measure the inside angle made by the lines and to determine how far outward the jaw projects.

Brow ridge: Note the presence or absence of a bony ridge above the eye sockets.

Teeth: Count the number of each kind of teeth in the lower jaw.

Procedure

PART A: APE SKULLS AND HUMAN SKULLS

1. Examine the diagrams of the skull and jaw of an ape and a human. Look for similarities and differences between the features listed in the chart "How to Interpret the Features of a Skull." Record your observations and measurements for each feature listed in Data Table 1.

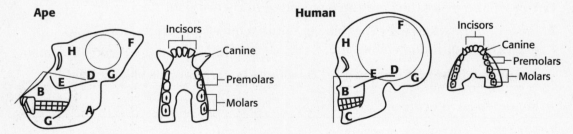

Sample data provided.

Data Table 1						
Name	**Cranial capacity (cm³)**	**Lower face area (cm²)**	**Brain area (cm²)**	**Jaw angle (degrees)**	**Brow ridge**	**Teeth**
Ape	422	420	230	130	Yes	6 molars, 4 premolars 2 canines, 4 inscisors
Human	1953	161	264	80	No	Same as above

Comparing Hominid Skulls *continued*

PART B: FOSSIL HOMINIDS

2. Examine the four fossil hominid skulls. On the hominid skulls, observe and measure four features that are listed in the chart "How to Interpret the Features of a Skull." Use the human skull as a model for taking measurements. Record your observations and measurements in Data Table 2.

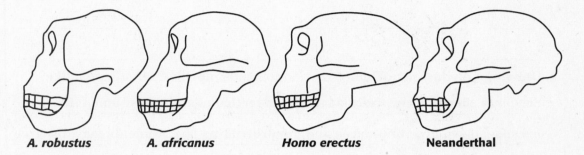

A. robustus   *A. africanus*   **Homo erectus**   **Neanderthal**

Sample data provided.

Data Table 2				
Name	**Lower face area (cm²)**	**Brain area (cm²)**	**Jaw angle (degrees)**	**Brow ridge**
Australopithecus robustus	300 (I)	260 (I)	115 (I)	Yes
Australopithecus africanus	252 (I)	190 (A)	125 (I)	Yes
Homo erectus	312 (I)	252 (I)	111 (I)	Yes
Neanderthal	300 (I)	308 (H)	115 (I)	Yes

3. Compare your data for the hominids with your data for the modern ape and human. Classify each feature of the hominid skulls as being apelike, humanlike, or intermediate by writing an *A*, *H*, or *I* next to your observation or measurement for that feature.

4. Using your data, predict the order in which the hominids shown here may have evolved.

Answers may vary. Most scientists think that *A africanus* evolved first,

followed by *A. robustus*, *H. erectus*, and Neanderthal.

Analyze and Conclude

1. **Summarizing Results** How did skull structure change as hominids evolved?

   As hominids evolved, the brow ridge was minimized, the brain area

   increased, and the jaw became less prominent.

2. **Drawing Conclusions** Which fossil skull is most apelike? most humanlike?

   Students should recognize that the Neanderthal skull is most humanlike

   because of its large brain area and nonprotruding jaw. Students may disagree

   about which hominid skull is most apelike, *A. robustus* or *A. africanus*. Both

   skulls have a small brain area, a brow ridge, and a protruding jaw.

3. **Further Inquiry** Write a new question about human evolution that could be
   explored with another investigation.

   Answers will vary. For example: What was the probable diet of fossil

   hominids based on their tooth and jaw structure?

Observing a Live Frog

Teacher Notes

TIME REQUIRED One 45-minute class period

SKILLS ACQUIRED
Collecting Data
Communicating
Inferring
Organizing and Analyzing Data

RATING

Easy $\longleftarrow$ 1   2   3   4 $\longrightarrow$ Hard

Teacher Prep–2
Student Setup–1
Concept Level–1
Cleanup–1

SCIENTIFIC METHODS

In this lab, students will:
Make Observations
Ask Questions
Analyze the Results
Draw Conclusions

MATERIALS

Materials for this lab can be ordered from WARD'S. Use the Lab Materials QuickList Software on the **One-Stop Planner CD-ROM** for catalog numbers and to create a customized list of materials for this lab. If you use frogs that are not native to your area, do not release them into the local environment when the investigation is completed. Indigenous frogs may be released into an appropriate habitat. Students may take indigenous frogs home to keep as pets if they have an appropriate habitat for them.

SAFETY CAUTIONS

Have students follow the guidelines from the National Association of Biology Teachers for handling vertebrate animals.

TIPS AND TRICKS

Provide each group with a frog in a small terrarium. Have groups take turns releasing their frog into the freshwater aquarium. After students complete their observations, remove each frog from the aquarium yourself to avoid injury to the frogs.

ANSWERS TO BEFORE YOU BEGIN

1. *amphibians*—animals that are adapted for living both on land and in water, usually referring to vertebrates in the class Amphibia; *nictitating membrane*—the extra eyelid of frogs' eyes; *tympanic membrane*—membrane located on a frog's head that receives vibrations and transmits them to the inner ear.

Exploration Lab

Observing a Live Frog

SKILLS

- Observing
- Relating

OBJECTIVES

- **Examine** the external features of a frog.
- **Observe** the behavior of a frog.
- **Explain** how a frog is adapted to life on land and in water.

MATERIALS

- live frog in a terrarium
- live insects (crickets or mealworms)
- 600 mL beaker
- aquarium half-filled with dechlorinated water

Before You Begin

Frogs, which are **amphibians,** are adapted for living on land and in water. For example, a frog's eyes have an extra eyelid called the **nictitating membrane.** This eyelid protects the eye when the frog is underwater and keeps the eye moist when the frog is on land. The smooth skin of a frog acts as a respiratory organ by exchanging oxygen and carbon dioxide with the air or water. The limbs of a frog enable it to move both on land and in water. In this lab, you will examine a live frog in both a terrestrial environment and an aquatic environment.

1. Write a definition for each boldface term in the paragraph above and for the following term: tympanic membrane. Use a separate sheet of paper.
   Answers appear in the Teacher's Notes.

2. You will be using the data table provided to record your data.

3. Based on the objectives for this lab, write a question you would like to explore about frogs.
   Answers will vary. For example: How are frogs adapted for living on land and

   in water?

Procedure
PART A: OBSERVING A FROG

1. Observe a live frog in a terrarium. Closely examine the external features of the frog. On a separate piece of paper, make a drawing of the frog. Label the eyes, nostrils, tympanic membranes, front legs, and hind legs.

Name _____ Class _____ Date _____

Observing a Live Frog *continued*

Data Table	
Behavior/structure	**Observations**
Breathing	
Eyes	
Legs	
Response to food	
Response to noise	
Skin	
Swimming behavior	

2. Watch the frog's movements as it breathes air into and out of its lungs. Record your observations.

3. Look closely at the frog's eyes, and note their location. Examine the upper and lower eyelids as well as a third transparent eyelid called a *nictitating membrane*. Describe how the eyelids move.

4. Study the frog's legs, and note the difference between the front and hind legs.

5. Place a live insect, such as a cricket or a mealworm, into the terrarium. Observe how the frog reacts.

6. Tap the side of the terrarium farthest from the frog, and observe the frog's response.

7. Place a 600 mL beaker in the terrarium. **CAUTION: Handle live frogs gently. Frogs are slippery! Do not allow a frog to injure itself by jumping from a lab table to the floor.** Carefully pick up the frog, and examine its skin. How does it feel? Now place the frog in the beaker. Cover the beaker with your hand, and carry it to a freshwater aquarium. Tilt the beaker, and gently lower it into the water until the frog swims out.

8. Watch the frog float and swim. Notice how the frog uses its legs to swim. Also notice the position of the frog's head. As the frog swims, bend down to view the underside of the frog. Then look down on the frog from above. Compare the color on the dorsal and ventral sides of the frog.

PART B: CLEANUP AND DISPOSAL

9. Dispose of broken glass in the designated waste containers. Put live animals in the designated containers. Do not pour chemicals down the drain or put lab materials in the trash unless your teacher tells you to do so.

10. Clean up your work area and all lab equipment. Return lab equipment to its proper place. Wash your hands thoroughly before you leave the lab and after you finish all work.

| Observing a Live Frog *continued*

Analyze and Conclude

1. Summarizing Information How does a frog use its hind legs for moving on land and in water?

A frog uses its long hind legs for jumping on land and for swimming in water.

2. Recognizing Relationships How does the position of a frog's eyes benefit the frog while it is swimming?

Having eyes on the top of its head enables a frog to see above the waterline

while the rest of the frog is submerged.

3. Analyzing Data What features of an adult frog provide evidence that it has an aquatic life and a terrestrial life?

aquatic life–webbed feet, eyes on top of head, respires through skin;

terrestrial life–jumping legs, breathes with lungs

4. Analyzing Methods Were you able to determine in this lab how a frog hears? Explain.

Answers will vary. Students should recognize that although they observed

the tympanic membrane, they cannot determine how a frog hears by

observing its actions.

5. Inferring Conclusions What can you infer about a frog's field of vision from the position of its eyes?

A frog can see in almost all directions.

6. Forming Hypotheses How is the coloration on the dorsal and ventral sides of a frog an adaptive advantage?

As viewed from above, the dark dorsal skin blends in with the color of the

water, making the frog less visible to predators from above. Viewed from

below, the light ventral skin blends in with the color of the sky, making the

frog less visible to aquatic predators.

7. Further Inquiry Write a new question about frogs that could be explored with another investigation.

Answers will vary. For example: How do two frogs interact on land and in

water?

Observing Color Change in Anoles

Teacher Notes

TIME REQUIRED Two 45-minute lab periods

SKILLS ACQUIRED

Observing   Collecting data
Designing experiments   Communicating

RATING

Teacher Prep–2
Student Setup–1   Easy ◄—— 1 —— 2 —— 3 —— 4 ——► Hard
Concept Level–2
Cleanup–1

SCIENTIFIC METHODS

In this lab, students will:

Make Observations   Analyze the Results
Ask Questions   Draw Conclusions
Test the Hypothesis

MATERIALS

Materials for this lab can be ordered from WARD'S. Use the Lab Materials QuickList Software on the **One-Stop Planner CD-ROM** for catalog numbers and to create a customized list of materials for this lab.

SAFETY CAUTIONS

Remind students to handle anoles with care. These lizards are easily frightened and can run quickly. If an anole is picked up by its tail, the tail may break off.

TIPS AND TRICKS

At least one hour before the lab exercise begins, place half the anoles in a terrarium lined with black construction paper. These anoles should turn brown. Place the remaining anoles in a terrarium lined with white construction paper. These anoles should turn green. Each terrarium should be illuminated by a fluorescent light and should contain a shallow dish of water.

Because anoles move quickly, you may want to place them in the jars for students in advance. Cover jars with lids that have air holes. Keep all jars in the same environment. Color change may be subtle and may be affected by feeding patterns or epinephrine (adrenaline) if the lizard is frightened. Very cool temperatures can induce a change to brown. Clean terrariums and jars after you have removed the anoles. Responsible and interested students may wish to adopt the anoles that are no longer needed in the classroom. Show students how to properly set up a vivarium to care for the anoles.

Exploration Lab

Observing Color Change in Anoles

SKILLS

- Using scientific methods
- Observing

OBJECTIVES

- **Observe** live anoles.
- **Relate** the color of an anole to the color of its surroundings.

MATERIALS

- glass-marking pencil
- 2 large, clear jars with wide mouths and lids with air holes
- 2 live anoles
- 6 shades each of brown and green construction paper, ranging from light to dark (2 swatches of each shade)

Before You Begin

Lizards are a group of **reptiles.** There are 250–300 species of anoles, lizards in the genus *Anolis.* Like chameleons, anoles can change color, ranging from brown to green. Anoles live in shrubs, grasses, and trees. Light level, temperature, and other factors, such as whether the animal is frightened or has eaten recently, can all affect the color of an anole. When anoles are frightened, they usually turn dark gray or brown and are unlikely to respond to other **stimuli.** Anoles generally change color within a few minutes. In this lab, you will observe the ability of anoles to change color when they are placed on different background colors. You will also determine how this ability might be an advantage to anoles.

1. Write a definition for each boldface term in the paragraph above.

reptiles—**ectothermic vertebrates with scaly, watertight skin; lungs; and a**

heart with partially or completely divided ventricle.

stimuli—**environmental factors that influence the behavior of an organism**

_____

2. You will be using the data table provided to record your data.

3. Based on the objectives for this lab, write a question you would like to explore about the color-changing behavior of anoles.

Answers will vary. For example: How quickly do anoles change color after

they move to a new background color?

Procedure
PART A: MAKE OBSERVATIONS

1. Observe live anoles in a terrarium. Make a list of characteristics that indicate that anoles are reptiles.

Easily observed characteristics include scaly skin and toes with claws.

_____

2. Work with a partner to place anoles to be studied in separate glass jars. **CAUTION: Handle anoles gently, and follow instructions carefully. Anoles run fast and are easily frightened. Plan your actions before you start.** By working efficiently, you can keep your anole from becoming overly frightened. Carefully pick up one anole by grasping it firmly but gently around the shoulders. Do not pick up anoles by their tail. Place the anole in a glass jar. Quickly and carefully place a lid with air holes on the jar.

3. When anoles become overly frightened, they remain dark. While you are designing your experiment, do not disturb your anoles, and let them recover from your handling.

PART B: DESIGN AN EXPERIMENT

4. Work with members of your lab group to explore one of the questions written for step 3 of **Before You Begin.** To explore the question, design an experiment that uses the materials listed for this lab.

> **You Choose**
>
> As you design your experiment, decide the following:
>
> **a.** what question you will explore
>
> **b.** what hypothesis you will test
>
> **c.** how many anoles you will need
>
> **d.** what background colors you will use
>
> **e.** how many times you will test each background with an anole
>
> **f.** how long you will observe each test and how you will keep track of time
>
> **g.** what your control will be
>
> **h.** what data to record in your data table

5. Write a procedure for your experiment. Make a list of all the safety precautions you will take. Have your teacher approve your procedure and safety precautions before you begin the experiment.

Sample data:

Data Table				
	Color 1 (green paper)		Color 2 (brown paper)	
Anole	Change	Time	Change	Time
1 (brown at start)	green	3 s	stayed brown	
2 (green at start)	stayed green		brown	3 s

6. Set up and carry out your experiment.

PART C: CLEANUP AND DISPOSAL

7.  Dispose of construction paper and broken glass in the designated waste containers. Put anoles in the designated container. Do not put lab materials in the trash unless your teacher tells you to do so.

8.  Clean up your work area and all lab equipment. Return lab equipment to its proper place. Wash your hands thoroughly before you leave the lab.

Analyze and Conclude

1. Summarizing Results Briefly state how the variable you tested influenced the color-changing behavior of anoles.

Students' answers should clearly describe how each test was conducted and should state their results in terms of color changes in the anoles. See Sample Data Table in the Teacher's Notes.

2. Evaluating Results Did any unplanned variables influence your data? (For example, was there a loud noise, or was a jar suddenly moved?)

Anoles may react to stimuli other than the independent variable selected. Students may list several uncontrolled variables, such as temperature or environmental stressors, which could have affected their results.

3. Analyzing Methods How could your experiment be modified to improve the certainty of your results?

Answers will depend on the groups' experimental designs. Students should suggest improving their methods by eliminating or controlling as many uncontrolled variables as possible.

Observing Color Change in Anoles *continued*

4. Analyzing Data Were there any inconsistencies in your data? (For example, two anoles reacted in different ways.) If so, offer an explanation for them.

Answers will vary. Students should include a possible explanation for any

inconsistency observed.

5. Drawing Conclusions After considering your data, make a statement about color-changing behavior in anoles.

Answers will vary. For example: Anoles change to a color that most closely

matches the color of their environment.

6. Further Inquiry Write a new question about anoles that could be explored with another investigation.

Answers will vary. For example: Would anoles react differently if there were

two in each jar?

Exploring Mammalian Characteristics

Teacher Notes
TIME REQUIRED 45 Minutes

SKILLS ACQUIRED
Classifying
Collecting Data
Inferring
Organizing and Analyzing Data

RATING
Easy ←——1——2——3——4——→ Hard

Teacher Prep–2
Student Setup–1
Concept Level–1
Cleanup–1

SCIENTIFIC METHODS

In this lab, students will:
Make Observations
Analyze the Results
Draw Conclusions

MATERIALS

Materials for this lab can be ordered from WARD'S. Use the Lab Materials QuickList Software on the **One-Stop Planner CD-ROM** for catalog numbers and to create a customized list of materials for this lab. Provide specimens, photos, or diagrams of a variety of mammalian and nonmammalian skulls. For handwashing, antibacterial soap is recommended.

SAFETY CAUTIONS

Review all safety symbols with students before beginning the lab. Advise students to handle specimens and laboratory equipment with care. Warn students not to touch their faces or eyes after handling samples.

TIPS AND TRICKS

Point out features of skulls that vary among animals, such as the position of the foramen magnum, the opening where the brain stem exits the skull. In bipedal animals, it is at the bottom of the skull, but in quadrupeds, it is at the back of the skull.

ANSWERS TO BEFORE YOU BEGIN

1. *mammal*—vertebrate that has hair, mammary glands, and specialized teeth

   hair—strands of protein that grow from the skin of mammals

   mammary glands—milk-producing glands of female mammals

   endothermy—the generation of one's own body heat by metabolism

   oil (sebaceous) glands—oil-producing glands located in the hair follicles of mammalian skin

   sweat glands—water-secreting glands in the skin of some mammals

Exploration Lab

Exploring Mammalian Characteristics

SKILLS

- Observing
- Drawing
- Inferring

OBJECTIVES

- **Examine** distinguishing characteristics of mammals.
- **Infer** the functions of mammalian structures.

MATERIALS

- hand lens or stereomicroscope
- prepared slide of mammalian skin
- compound microscope
- mirror
- specimens or pictures of vertebrate skulls
  (some mammalian, some nonmammalian)

Before You Begin

Mammals are vertebrates with **hair, mammary glands,** a single lower jawbone, and specialized teeth. Other characteristics of mammals include **endothermy** and a four-chambered heart. Mammals also have **oil (sebaceous) glands** in their skin, and most have **sweat glands.** In this lab, you will examine some of the characteristics of mammals that distinguish them from other vertebrates.

1. Write a definition for each boldface term in the paragraph above. Use a separate sheet of paper. **Answers appear in the Teacher's Notes.**

2. Record your data in the data table provided.

3. Based on the objectives for this lab, write a question you would like to explore about the characteristics of mammals.

   <u>Answers will vary. For example: How do the teeth of humans compare to</u>

   <u>those of other mammals?</u>

Exploring Mammalian Characteristics *continued*

Procedure
PART A: EXAMINING MAMMALIAN SKIN

1. Use a hand lens to look at several areas of your skin, including areas that appear to be hairless. Record your observations.

   Students should note that they can see pores and hairs except on the

   fingers, palms, toes, and soles of the feet.

2. Look at a prepared slide of mammalian skin under low power of a compound microscope. Notice the glands in the skin. Look for the oil (sebaceous) glands and the sweat glands. Draw and label an example of each type of gland. Use a separate sheet of paper. **Student's drawings should show oil and sweat glands.**

PART B: EXAMINING MAMMALIAN TEETH AND SKULLS

3. Wash your hands thoroughly with soap and water. Use a mirror to look in your mouth. Identify the four kinds of mammalian teeth you see.

   Students should identify incisors, canines, premolars, and molars.

4. Count each kind of tooth on one side of your lower jaw. Multiply the number of each kind of tooth by 4, and record these numbers in the appropriate columns of the data table below. Wash your hands again before continuing. **Sample data is shown in the table below.**

5. Look at the skulls of several mammals. Identify the kinds of teeth in each skull. For each skull, find the number of each kind of tooth as you did in step 4. **The number and type of teeth will vary according to the animal.**

Sample Data

Data Table				
Mammal	**Incisors**	**Canines**	**Premolars**	**Molars**
Human	$2 \times 4 = 8$	$1 \times 4 = 4$	$2 \times 4 = 8$	$3 \times 4 = 12$

6. Look at the skulls of several nonmammalian vertebrates, and compare nonmammalian teeth to mammalian teeth.

   Answers will vary. Nonmammalian teeth will differ significantly from

   mammalian teeth.

| Exploring Mammalian Characteristics *continued*

7. Compare the jaws of mammalian skulls to those of nonmammalian vertebrates. As you look at each skull, notice the structure of the lower jawbone and how the upper jawbone and the lower jawbone connect.

PART C: CLEANUP AND DISPOSAL

8. Dispose of broken glass in the waste container designated by your teacher.

9. Clean up your work area and all lab equipment. Return lab equipment to its proper place. Wash your hands thoroughly before you leave the lab and after you finish all work.

Analyze and Conclude

1. Summarizing Information List the characteristics that distinguish mammals from other vertebrates.

Hair, single jawbone, specialized teeth, and mammary glands
_____

2. Interpreting Graphics Compare the amount of hair on humans to that on a skunk and a dolphin.

Humans have far less hair than skunks have and more hair than dolphins

have.

3. Inferring Relationships What role might hair or fur play in enabling mammals to be endotherms?

Hair serves as insulation from extreme external temperatures and helps

maintain a constant internal temperature.

4. Forming Hypotheses Besides the role of hair you identified in item 3 above, what other roles do you think hair might play in mammals?

Answers will vary. Hair enables a mammal to gather sensory information

from its surroundings. It also helps to protect a mammal's skin from water

loss, sunburn, and injury.

5. Recognizing Patterns Where are the oil (sebaceous) glands located in the skin of mammals?

Sebaceous glands are located inside the hair follicles.

Exploring Mammalian Characteristics *continued*

6. Forming Hypotheses Do you think skunks and dolphins have more sweat glands or fewer sweat glands than humans have? Explain.

Answers will vary. Students should recognize that skunks, as well as

dolphins, are not likely to have as many sweat glands as humans. Humans

have very little hair for insulation. Sweat glands help to cool the body.

Skunks have sufficient hair for efficient insulation. Dolphins live in water

and have no need for sweat glands.

7. Comparing Structures How is the mammalian jaw different from nonmammalian jaws?

Mammals have a single jawbone that connects to the skull at a joint.

8. Inferring Conclusions Based on the shape of your teeth, would you classify humans as carnivores (meat eaters), herbivores (plant eaters), or omnivores (meat and plant eaters)? Explain.

Answers may vary. Students should recognize that humans are omnivores.

Humans have sharp incisors and canines for tearing meat, and flat molars

and premolars for grinding plant material.

9. Evaluating Conclusions Justify the following conclusion: The kinds and shapes of a mammal's teeth can be used to determine its diet.

Answers will vary. Carnivores have long, sharp canines for grabbing prey and

tearing meat. Herbivores have flattened teeth for grinding and crushing

plant material.

10. Further Inquiry Write a new question about the characteristics of mammals that could be explored with a new investigation.

Answers will vary. For example: How is a mammal's muzzle shape related to

its diet?

Studying Nonverbal Communication

Teacher Notes

TIME REQUIRED 45 minutes of classroom time; 30 minutes of time outside of class

SKILLS ACQUIRED

Identifying/Recognizing Patterns
Collecting Data
Organizing and Analyzing Data
Interpreting

RATING

Easy ◄——— 1   2   3   4 ———► Hard

Teacher Prep–1
Student Setup–1
Concept Level–2
Cleanup–1

SCIENTIFIC METHODS

In this lab, students will:
Make Observations
Analyze the Results
Draw Conclusions

MATERIALS AND EQUIPMENT

Students will need a stopwatch or a clock with a second hand, paper, and a pencil.

SAFETY CAUTIONS

Caution students to observe people in a safe setting and in an unobtrusive way that cannot be construed as offensive. Behavior changes quickly.

TIPS AND TRICKS

Form groups of two or three students. It is important that members of each group have time together outside of class to conduct their observations. Each observation will require the presence of at least two members of the group. Suggest that at least one student of the pair or group observe the behavior while another student records observations.

ANSWERS TO BEFORE YOU BEGIN

1. *posture*—body position; *stance*—position of the body while standing; *equal stance*—stance in which the body weight is supported equally by both legs; *unequal stance*—stance in which more of the body weight is supported by one leg than by the other

Exploration Lab

Studying Nonverbal Communication

SKILLS

- Observing
- Analyzing
- Graphing

OBJECTIVES

- **Recognize** that posture is a type of nonverbal communication.
- **Observe** how human posture changes during a conversation.
- **Determine** the relationship of gender to the postural changes that occur during a conversation.

MATERIALS

- stopwatch or clock with a second hand
- paper
- pencil

Before You Begin

People communicate nonverbally with their **posture,** or body position. The position of the body while standing is called the **stance.** In an **equal stance,** the body weight is supported equally by both legs. In an **unequal stance,** more weight is supported by one leg than by the other. In this lab, you will observe and analyze how stance changes during conversations between pairs of people who are standing.

1. Write a definition for each boldface term in the paragraph above. Use a separate sheet of paper. **Answers appear in the Teacher's Notes.**

2. You will be using the data table provided to record your data. Notice that the second and third tables are continuations of the first. Use a separate sheet of paper if you need to make new data tables for longer conversations.

3. Based on the objectives for this lab, write a question you would like to explore about nonverbal communication.

   Answers will vary. For example: What effects do conflicting postures have on

   an observer?

| **Studying Nonverbal Communication** *continued* |

Procedure

PART A: OBSERVING BEHAVIOR

1. Work in a group of two or three to observe conversations between pairs of people. Each conversation must last between 45 seconds and 5 minutes. One person in your group should be the timekeeper and the other group members should record data. **Note:** Be sure that your subjects are unaware they are being observed.

Data Table					
Pairs	**Gender Involved**	**Gender Observed**	**15 second intervals**		
			15 s	**30 s**	**45 s**
1					
2					
3					

Data Table					
Pairs	**Gender Involved**	**Gender Observed**	**15 second intervals**		
			1 min	**1 min 15 s**	**1 min 30 s**
1					
2					
3					

Data Table					
Pairs	**Gender Involved**	**Gender Observed**	**15 second intervals**		
			1 min 45 s	**2 min**	**2 min 15 s**
1					
2					
3					

Observations in steps 2–6 will vary. Check that students have correctly entered all observations in their data table.

2. Observe at least three conversations. Record the genders of the two participants in each conversation and the gender of the one person whose posture you observe. **Note:** Be sure that the timekeeper accurately clocks the passage of each 15-second interval.

3. For each 15-second interval, record all of the changes in stance by the person you are observing. For example, note every time your subject shifts from an equal stance to an unequal stance, or vice versa. To record the stance simply, you may write *E* to identify an equal stance and *U* to identify an unequal stance.

Studying Nonverbal Communication *continued*

4. If the subject assumes an unequal stance, also record the number of weight shifts from one foot to the other. Indicate a weight shift simply by writing *W*.

5. When a conversation ends, write down whether the pair departed together or separately. To record this, write *T* to indicate departing together or *S* to indicate departing separately.

6. After you have completed each observation, tally the total number of weight shifts within each 15-second block. **IMPORTANT!** Retain data only for conversations that last at least 45 seconds.

If a conversation ends before you have collected data for 45 seconds, observe another conversation.

PART B: ANALYZING DATA

7. After all observations have been completed, combine the data from all of the groups in your class. Analyze the data, without regard to gender.

a. Determine the most common stance during the first 15 seconds of a conversation, the middle 15 seconds, and the last 15 seconds. Make a bar graph on a separate sheet of paper to summarize the class data.

Answers will vary. Students may find that the most common stance is with

equal weight on each leg. Check that the bar graphs reflect the combined

class data.

b. Find the average number of weight shifts in the beginning, middle, and end intervals. Make a bar graph on a separate sheet of paper to summarize the class data.

Answers will vary. Students may find that the number of weight shifts

increased toward the end of intervals. Check that the bar graphs reflect

the combined class data.

8. Repeat step 7, but analyze the data according to gender this time.

Answers will vary. Check that students have correctly analyzed the combined

class data by gender. The bar graphs should reflect the analysis by gender.

9. Compile the data and make bar graphs on a separate sheet of paper for each of the following: males talking with a male, males talking with a female, females talking with a male, and females talking with a female. Compare these graphs with the ones you made in step 7.

Answers will vary. Check that students have correctly compiled the data. The

bar graphs should reflect the compiled data.

| **Studying Nonverbal Communication** *continued*

Analyze and Conclude

1. Analyzing Results Which stance was used most often during a conversation?

Answers will vary. Students will most likely find that the most common

stance is with equal weight on each leg.

2. Recognizing Relationships Which behavior most often signals that a conversation is about to end: stance change or weight shift?

Answers will vary. Students should support their answers by citing their

data.

3. Drawing Conclusions Do males and females differ in their departure signals? Justify your conclusion.

While researchers have not found significant differences between the sexes

in these signals, your students may see some trends due to the small sample

size. This could lead to an interesting discussion about statistics.

4. Forming Hypotheses What do you think might be an adaptive significance of a departure signal?

Answers will vary, but students may suggest that nonverbal cues may make

the departure expected and less abrupt. This may ensure that the person

leaving does not insult the other person.

5. Forming Reasoned Opinions What other behaviors you observed were forms of nonverbal communication? Justify your answer.

Answers will vary, but students may mention facial expressions and head,

hand, arm, or eye movements.

6. Further Inquiry Write a new question about animal behavior that could be explored with a new investigation.

Answers will vary. For example: What postural changes occur when two dogs

greet each other? How does nonverbal communication differ among different

cultures?

Analyzing the Work of Muscles

Teacher Notes

TIME REQUIRED One 50-minute class period.

SKILLS ACQUIRED
Using Scientific Methods
Data Collection
Data Interpretation

RATING

Easy ◄——— 1   2   3   4 ———► Hard

Teacher Prep–1
Student Setup–1
Concept Level–1
Cleanup–1

SCIENTIFIC METHODS

In this lab, students will:
Make Observations
Analyze the Results
Draw Conclusions
Communicate the Results

MATERIALS

Materials for this lab can be ordered from WARD'S. Use the Lab Materials
QuickList Software on the **One-Stop Planner CD-ROM** for catalog numbers
and to create a customized list of materials for this lab.

SAFETY CAUTION

Any student feeling pain from the exercise should stop exercising immediately.

TIPS AND TRICKS

Some students' hands may not be large enough to grip the hand grips. Pair
students who have small hands and give them a tennis ball to squeeze.

Exploration Lab

Analyzing the Work of Muscles

SKILLS

- Using scientific methods
- Data collection
- Data interpretation

OBJECTIVE

- **Relate** muscles to the work they do.
- **Observe** the effects of fatigue

MATERIALS

- watch with second hand
- graph paper
- spring hand grips

Before You Begin

Muscles are attached to bones. As muscles contract, they move the bones to which they are attached. This is a basic type of work accomplished by the human body. As muscles are used, lactic acid builds up, resulting in **fatigue.** In this lab you will investigate how fatigue affects the amount of work that muscles can do.

1. Write a definition for the bold face term in the preceding paragraph.

Fatigue **means "to become tired."**

2. Use the data table below.

Data Table For Muscle Contractions														
Number of Muscle Contractions in 10-Second Intervals														
Time	1st 10 sec	2nd 10 sec	3rd 10 sec	4th 10 sec	5th 10 sec	6th 10 sec	7th 10 sec	8th 10 sec	9th 10 sec	10th 10 sec	11th 10 sec	12th 10 sec	13th 10 sec	14th 10 sec
Trial 1														
Trial 2														
Trial 3														
Trial 4														

PROCEDURE

1. Perform this investigation with a partner. Designate one laboratory partner to observe and record while the other performs the experiment.

2. Hold the spring hand grips in your left hand if you are right-handed, or in your right hand if you are left-handed. Squeeze the grips rapidly and as hard as possible at a steady pace, until complete fatigue is experienced in the muscles of your hand and forearm.

3. The recorder should count and record the number of squeezes for every 10 seconds.

4. Allow the experimenter to rest for one minute and repeat the procedure for two more trials.

5. Record the data in your table. Some spaces may be left blank.

6. Switch roles with your partner and repeat the procedure.

Analyze and Conclude

1. **Summarizing Results** Use graph paper to plot the results of the three trials on a graph. The *x*-axis should be used for time in seconds and the *y*-axis for the number of muscle contractions. **Student graphs will vary, but the number of muscle contractions should decrease over time.**

2. **Analyzing Data** Account for the differences in the amount of work done by the muscles during the three trials.

   When muscles are fatigued, they can do less work than when they are not

   fatigued.

3. **Drawing Conclusions** What is the relationship between the work muscles can do and fatigue?

   The muscles have more energy, in the form of ATP, when they are first used,

   but the supply is quickly used and lactic acid is formed as a waste product.

   Time is needed for ATP molecules to re-form and lactic acid to dissipate.

4. **Predicting Patterns** How does the work done in the muscles of your hands and arms relate to the work done by the muscle of your heart?

   Both types of work involve muscles contracting. However, the heart muscle is

   able to contract regularly without fatigue because it rests between beats.

5. **Further Inquiry** Compare the charts and the graphs of the athletes and nonathletes.

   Answers may vary, but athletes may have muscles that show fatigue less

   quickly than nonatheletes.

Determining Lung Capacity

Teacher Notes

TIME REQUIRED 30–40 minutes

SKILLS ACQUIRED
Collecting Data
Measuring
Organizing and Analyzing Data

RATING

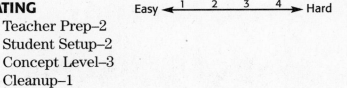

Teacher Prep–2
Student Setup–2
Concept Level–3
Cleanup–1

SCIENTIFIC METHODS

In this lab, students will:
Make Observations
Analyze the Results
Draw Conclusions

MATERIALS

Materials for this lab can be ordered from WARD'S. Use the Lab Materials QuickList Software on the **One-Stop Planner CD-ROM** for catalog numbers and to create a customized list of materials for this lab.

SAFETY CAUTIONS

Discuss all safety symbols and caution statements with students. Direct students not to share spirometer mouthpieces. Ask if any students have any respiratory conditions (including illnesses). These students should not participate in this exercise as subjects. They can participate as observers. Make sure students discard used spirometer mouthpieces in the designated container.

TIPS AND TRICKS

Place enough materials for a class on a supply table. Provide plenty of clean spirometer mouthpieces. Set up a labeled container for disposal of spirometer mouthpieces. Review the use of the spirometer before students begin the lab. You may wish to have students record their data in a class data table on the board. They need not identify themselves by name, only by gender, noting whether they are an athlete or nonathlete, and a smoker or nonsmoker. Calculate means for males, females, smokers, athletes, nonathletes, and nonsmokers.

ANSWERS TO BEFORE YOU BEGIN

1. *lung capacity*—the amount of air that can be inhaled and exhaled by the lungs; *tidal volume*—the volume of air breathed into or out of the lungs in a normal breath; *inspiratory reserve volume*—the volume of air that can be forcefully inhaled after a normal inhalation; *expiratory reserve volume*—the volume of air that can be forcefully exhaled after a normal exhalation; *vital capacity*—the maximum volume of air that can be inhaled or exhaled; *residual volume*—the volume of air that remains in the lungs after you have exhaled all the air that you can; *spirometer*—an instrument used to measure the volume of air exhaled from the lungs

Name _____ Class _____ Date _____

Determining Lung Capacity

SKILLS

- Measuring
- Organizing data
- Comparing

OBJECTIVES

- **Measure** your tidal volume, vital capacity, and expiratory reserve volume.
- **Determine** your inspiratory reserve capacity and lung capacity.
- **Predict** how exercise will affect tidal volume, vital capacity, and lung capacity.

MATERIALS

- spirometer
- spirometer mouthpiece

Before You Begin

Lung capacity is the total volume of air that the lungs can hold. The lung capacity of an individual is influenced by many factors, such as gender, age, strength of diaphragm and chest muscles, and disease.

During normal breathing, only a small percentage of your lung capacity is inhaled and exhaled. The amount of air inhaled or exhaled in a normal breath is called the **tidal volume.** An additional amount of air, called the **inspiratory reserve volume,** can be forcefully inhaled after a normal inhalation. The **expiratory reserve volume** is the amount of air that can be forcefully exhaled after a normal exhalation. **Vital capacity** is the maximum amount of air that can be inhaled or exhaled. Even after you have exhaled all the air you can, a significant amount of air called the **residual volume** still remains in your lungs.

In this lab, you will determine your lung capacity by using a **spirometer,** which is an instrument used to measure the volume of air exhaled from the lungs.

1. Write a definition for each boldface term in the paragraph above. Use a separate sheet of paper. **Answers appear in the Teacher's Notes.**

2. You will be using the data table provided to record your data.

3. Based on the objectives for this lab, write a question about breathing that you would like to explore.

   Answers will vary. For example: What is my vital capacity and lung capacity?

   _____

   _____

Determining Lung Capacity *continued*

Procedure

PART A: MEASURING VOLUME

1. Place a clean mouthpiece in the end of a spirometer. **CAUTION: Many diseases are spread by body fluids, such as saliva. Do NOT share a spirometer mouthpiece with anyone.**

2. To measure your tidal volume, first inhale a normal breath. Then exhale a normal breath into the spirometer through the mouthpiece. Record the volume of air exhaled in the data table below.

   Sample data:

Data Table	
Tidal volume	**500 mL**
Expiratory reserve volume	**100 mL**
Inspiratory reserve volume	**4,000 mL**
Vital capacity	**4,600 mL**
Estimated residual volume	**1,200 mL**
Estimated lung capacity	**6,000 mL**

3. To measure your expiratory reserve volume, first inhale a normal breath and then exhale normally. Then forcefully exhale as much air as possible into the spirometer. Record this volume.

4. To measure your vital capacity, first inhale as much air as you can, and then forcefully exhale as much air as you can into the spirometer. Record this volume.

PART B: CALCULATING LUNG CAPACITY

The table below contains average values for residual volumes and lung capacities for young adults.

Residual Volumes and Lung Capacities		
	Males	**Females**
Residual volume*	1,200 mL	900 mL
Lung capacity*	6,000 mL	4,500 mL

*Athletes can have volumes 30–40% greater than the average for their gender.

5. Inspiratory reserve volume (IRV) can be calculated by subtracting tidal volume (TV) and expiratory reserve volume (ERV) from vital capacity (VC). The formula for this calculation is as follows:

$$IRV = VC - TV - ERV$$

Use the data in the data table and the equation above to calculate your estimated inspiratory reserve volume.

6. Lung capacity (LC) can be calculated by adding residual volume (RV) to vital capacity (VC). The formula for this calculation is as follows:

$$LC = VC + RV$$

Use the data in the data table and the table above to calculate your estimated lung capacity.

PART C: CLEANUP AND DISPOSAL

7. Dispose of your mouthpiece in the designated waste container.

8. Clean up your work area and all lab equipment. Return lab equipment to its proper place. Wash your hands thoroughly before you leave the lab and after you finish all work.

Analyze and Conclude

1. Interpreting Data How does your expiratory reserve volume compare with your inspiratory reserve volume?

Answers will vary. The expiratory reserve volume should be less than half

the inspiratory reserve volume.

2. Interpreting Tables How does the residual volume and lung capacity of an average young adult female compare with those of an average young adult male?

The average value for females is less than the average value for males.

3. Analyzing Data How did your tidal volume compare with that of others?

Answers will vary.

4. Recognizing Relationships Why was the value you found for your lung capacity an estimated value?

Answers will vary. The value for residual volume cannot be measured with a

spirometer. The figure given for residual volume is an average value, which

is probably not the same as a student's actual value. Therefore, the lung

capacity found, which includes the residual volume, is also an estimate.

5. Analyzing Methods Why didn't you measure inspiratory reserve volume directly?

Answers will vary. Students should state that there would be no way to judge

when they had exhaled the amount of air inhaled after a normal inhalation

because they must breathe out to take a reading.

6. Inferring Conclusions Why would males and athletes have greater vital capacities than females?

Answers may vary. Vital capacity depends mostly on body size. Respiratory

muscle strength, lung expansion capacity, and chest size are also important.

Males are usually larger than females. Athletes have stronger respiratory

muscles and greater lung expansion capacity.

7. Justifying Conclusions Use data from your class to justify the conclusion that exercise increases lung capacity.

Answers will vary. Students should see that classmates who exercise

regularly have a greater lung capacity than those that do not.

8. Further Inquiry Write a new question that could be explored with another investigation.

Answers will vary. For example: How do asthma and emphysema reduce the

efficiency of the respiratory system as measured by a spirometer?

Demonstrating Lactose Digestion

Teacher Notes

TIME REQUIRED Two 45-minute periods

SKILLS ACQUIRED
Collecting Data
Designing Experiments
Experimenting
Inferring
Organizing and Analyzing Data

RATING
Easy ◄——$\overset{1}{\quad}$——$\overset{2}{\quad}$——$\overset{3}{\quad}$——$\overset{4}{\quad}$——► Hard

Teacher Prep–2
Student Setup–3
Concept Level–4
Cleanup–1

SCIENTIFIC METHODS
In this lab, students will:
Make Observations
Ask Questions
Test the Hypothesis
Analyze the Results
Draw Conclusions

MATERIALS

Materials for this lab can be ordered from WARD'S. Use the Lab Materials QuickList Software on the **One-Stop Planner CD-ROM** for catalog numbers and to create a customized list of materials for this lab. All milk products can be poured down the drain. Glucose test strips can be placed in the trash.

SAFETY CAUTIONS

Discuss all safety symbols and caution statements with students.

TIPS AND TRICKS

Glucose test strips and commercial milk-treatment products that contain yeast-derived lactase are available from most pharmacies. Buy several boxes of the glucose test strips so the lab groups each have a color guide. Avoid buying the yeast-derived lactase in solid form. Do not use milk-treatment product tablets. The tablets will give positive glucose results. Each lab group will need copies of the information sheet that accompanies the product. When you read the students' protocols for approval, direct students to use only drops of milk and milk-treatment product instead of full quarts of milk. Glucose test strips are more convenient than Benedict's test because they do not require a hot water bath and they give faster results.

Before students begin the lab ask how you could tell if a milk-treatment product digests lactose. The presence of glucose and/or galactose in the milk after addition of the treatment product would indicate that lactose has been broken down. Some students will be unsure about what controls to use. Instruct students to test the glucose test strip for a positive glucose test in a known glucose solution.

ANSWERS TO BEFORE YOU BEGIN

1. *lactose intolerance*—an inability to digest the lactose in milk or milk-containing products due to the lack of the enzyme lactase

   lactose—a disaccharide found in milk and composed of a glucose unit and galactose unit

   digestion—the breakdown of food molecules into simpler molecules that can be taken up by cells

   lactase—the enzyme that catalyzes the breakdown of lactose

Exploration Lab

Demonstrating Lactose Digestion

SKILLS

- Using scientific methods
- Observing
- Comparing

OBJECTIVES

- **Describe** the relationship between enzymes and the digestion of food molecules.
- **Evaluate** the ability of a milk-treatment product to promote lactose digestion.
- **Infer** the presence of lactose in milk and foods that contain milk.

MATERIALS

- milk-treatment product (liquid)
- toothpicks
- depression slides
- droppers
- whole milk
- glucose solution
- glucose test strips

SAFETY

 CAUTION: Always wear safety goggles and a lab apron to protect your eyes and clothing.

CAUTION: Do not touch or taste any chemicals. Know the location of the emergency shower and eyewash station and how to use them. If you get a chemical on your skin or clothing, wash it off at the sink while calling to the teacher. Notify the teacher of a spill. Spills should be cleaned up promptly, according to your teacher's directions.

CAUTION: Glassware is fragile. Notify the teacher of broken glass or cuts. Do not clean up broken glass or spills with broken glass unless the teacher tells you to do so.

Before You Begin

People with a condition known as **lactose intolerance** often experience stomach and intestinal pain, bloating, and diarrhea when they eat foods that contain milk. These symptoms result from an inability to digest lactose, a sugar found in milk. **Lactose** is a disaccharide made of one glucose unit and one galactose unit. Lactose molecules are broken down into glucose and galactose molecules during **digestion.** People who cannot digest lactose do not produce **lactase,** the digestive enzyme that aids the breakdown of lactose. In this lab, you will investigate a milk-treatment product that is designed to aid lactose digestion.

1. Write a definition for each boldface term in the paragraph above. Use a separate sheet of paper. **Answers appear in the Teacher's Notes.**

2. List at least 10 foods that contain milk.

   Answers may vary. Examples may include milk, cheese, butter, ice cream, and

   cottage cheese.

3. You will be using the data table provided to record your data.

4. Based on the objectives for this lab, write a question you would like to explore about enzymes and digestion.

   Answers may vary. For example: How effective is a milk-treatment product in

   breaking down lactose?

Data Table		
Solution	**Result (+ or −)**	**Interpretation**

Procedure

PART A: DESIGN AN EXPERIMENT

1. Read the information sheet that comes with the milk-treatment product. Discuss with your lab group what the product is and what it does. Write a summary of your discussion for your lab report. Use a separate sheet of paper.

2. Work with the members of your lab group to explore one of the questions written for step 4 of **Before You Begin.** To explore the question, design an experiment that uses the materials listed for this lab.

> **You Choose**
>
> As you design your experiment, decide the following:
>
>   **a.** what question you will explore
>
>   **b.** what hypothesis you will test
>
>   **c.** what your controls will be
>
>   **d.** how much milk and milk-treatment product to use for each test
>
>   **e.** how to determine whether lactose was broken down
>
>   **f.** what data to record in your data table

3. Write the procedure for your group's experiment. Make a list of all the safety precautions you will take. Have your teacher approve your procedure and safety precautions before you begin the experiment.

4. ◆ ◆ ◆ ◆ Set up your group's experiment, and collect data.

PART B: CLEANUP AND DISPOSAL

5. ◆ Dispose of solutions, broken glass, and glucose test strips in the designated waste containers. Do not pour chemicals down the drain or put lab materials in the trash unless your teacher tells you to do so.

6. ◆ Clean up your work area and all lab equipment. Return lab equipment to its proper place. Wash your hands thoroughly before you leave the lab and after you finish all work.

Analyze and Conclude

1. Summarizing Information What are the milk-treatment product's ingredients?

yeast-derived lactase
_____

Demonstrating Lactose Digestion *continued*

2. Recognizing Relationships What is the relationship between lactose and lactase?

Lactose is a disaccharide found in milk. Lactase is an enzyme that cat-

alyzes the breakdown of lactose into glucose and galactose. Lactose is

the substrate for the enzyme lactase.

3. Analyzing Methods What role did the glucose solution play in your experiment?

The glucose solution is a positive control for the experiment.

4. Drawing Conclusions What does the milk-treatment product do to milk?

It breaks down the lactose in the milk into glucose and galactose.

5. Analyzing Conclusions How do your results justify your conclusion?

Answers may vary. Students indirectly tested for the breakdown of lactose

by testing for the presence of glucose. Milk that was treated with the milk-

treatment product tested positive for glucose, while milk that was not

treated was negative for glucose.

6. Evaluating Methods Why should you test the milk-treatment product with glucose test strips?

to eliminate the milk-treatment product as a possible source of glucose

7. Analyzing Results What do you infer from the results of this lab about treatments for other medical problems resulting from enzyme deficiencies?

Other disorders caused by enzyme deficiencies might be treated with

products that contain the missing enzymes.

8. Forming Reasoned Opinions As a person grows older, will he or she be more likely or less likely to develop lactose intolerance? Explain your answer.

Answers may vary. The risk of lactose intolerance increases with age because

lactase is required most during infancy and early childhood when milk is a

major part of the diet.

Demonstrating Lactose Digestion *continued*

9. **Predicting Patterns** Do you think lactose intolerance might be inherited? Explain your answer.

Answers may vary. As an enzyme, lactase is a protein and proteins are

encoded in DNA, the molecule responsible for inheritance. Certain

populations have a higher incidence of lactose intolerance than others.

10. **Further Inquiry** Write a new question about enzymes and digestion that could be explored with another investigation.

Answers may vary. For example: How would blood glucose levels compare

in people eating treated dairy products with those eating untreated dairy

products?

Simulating Disease Transmission

Teacher Notes

TIME REQUIRED 30 minutes

SKILLS ACQUIRED
Collecting Data
Communicating
Identifying/Recognizing Patterns
Interpreting
Organizing and Analyzing Data

RATING
Easy ← 1   2   3   4 → Hard

Teacher Prep–3
Student Setup–3
Concept Level–2
Cleanup–3

SCIENTIFIC METHODS

In this lab, students will:
Make Observations
Analyze the Results
Draw Conclusions
Communicate the Results

MATERIALS

Materials for this lab can be ordered from WARD'S. Use the Lab Materials QuickList Software on the **One-Stop Planner CD-ROM** for catalog numbers and to create a customized list of materials for this lab.

SAFETY CAUTIONS

Review all safety symbols with students before beginning the lab.

TIPS AND TRICKS

Before students begin, remind them of safety practices, and emphasize that they should not allow solutions to touch their skin or clothes. Use stock dropper bottles of distilled water for half the students and dropper bottles containing 10 percent ascorbic acid for the rest of the students. Both solutions will be clear. A 10 percent solution of ascorbic acid can be prepared by dissolving 10 g of ascorbic acid in 50 mL of water and diluting it to 100mL with water. Each student will need a dropper bottle of unknown solution, a large test tube, and indophenol indicator solution. If the procedure is followed precisely, the route of transmission can be traced easily. Be sure to warn students not to select a new partner until each round is over and all students have exchanged liquids for that round. If one student makes a mistake, it can make it very difficult to trace the route of transmission.

ANSWERS TO BEFORE YOU BEGIN

1. *communicable disease*—a disease that can be transmitted from one person to another

   pathogen—a disease-causing agent such as a bacterium or a virus

   aerosol—a mist of tiny droplets of a liquid

Skills Practice Lab

Simulating Disease Transmission

SKILLS

- Modeling
- Organizing and analyzing data

OBJECTIVES

- **Simulate** the transmission of a disease.
- **Determine** the original carrier of the disease.

MATERIALS

- safety goggles
- lab apron
- disposable gloves
- dropper bottle of unknown solution
- large test tube
- indophenol indicator

SAFETY

 CAUTION: Always wear safety goggles and a lab apron to protect your eyes and clothing.

CAUTION: Do not touch or taste any chemicals. Know the location of the emergency shower and eyewash station and how to use them. If you get a chemical on your skin or clothing, wash it off at the sink while calling to the teacher. Notify the teacher of a spill. Spills should be cleaned up promptly, according to your teacher's directions.

CAUTION: Glassware is fragile. Notify the teacher of broken glass or cuts. Do not clean up broken glass or spills with broken glass unless the teacher tells you to do so.

Before You Begin

Communicable diseases are caused by **pathogens** and can be transmitted from one person to another. You can become infected by a pathogen in several ways, including by drinking contaminated water, eating contaminated foods, receiving contaminated blood, and inhaling infectious **aerosols** (droplets from coughs or sneezes). In this lab, you will simulate the transmission of a communicable disease. After the simulation, you will try to identify the original infected person in the closed class population.

| Simulating Disease Transmission *continued*

1. Write a definition for each boldface term in the paragraph above. Use a separate sheet of paper. **Answers appear in the Teacher's Notes.**

2. You will be using the data table provided to record your data.

3. Based on the objectives for this lab, write a question you would like to explore about disease transmission.

Answers will vary. Sample answer: Can you determine whether a disease has

been caused by the passing of pathogens from person to person or by

environmental conditions?

Procedure

PART A: SIMULATE DISEASE TRANSMISSION

1. ◆ ◆ ◆ Put on safety goggles, a lab apron, and gloves.

2. You will be given a dropper bottle of an unknown solution. When your teacher says to begin, transfer 3 dropperfuls of your solution to a clean test tube.

3. Select a partner for Round 1. Record the name of this partner in Data Table 1.

Data Table 1	
Round number	**Partner's name**

4. Pour the contents of one of your test tubes into the other test tube. Then pour half the solution back into the first test tube. You and your partner now share any pathogens either of you might have.

5. On your teacher's signal, select a new partner for Round 2. Record this partner's name in Data Table 1. Repeat step 4.

6. On your teacher's signal, select another new partner for Round 3. Record this partner's name. Repeat step 4.

7. Add one dropperful of indophenol indicator to your test tube. "Infected" solutions will stay colorless or turn light pink. "Uninfected" solutions will turn blue. Record the results of your test. **Most solutions will be "infected" and will remain colorless or turn pink; the remaining "uninfected" solutions will turn blue. Each repetition of the "sharing" activity increased the number of "infected" individuals.**

Simulating Disease Transmission *continued*

PART B: TRACE THE DISEASE SOURCE

8. If you are infected, write your name and the name of your partner in each round on the board or on an overhead projector. Mark your infected partners. Record all the data for your class in Data Table 2.

Data Table 2			
Name of infected person	**Names of infected person's partners**		
	Round 1	**Round 2**	**Round 3**

9. To trace the source of the infection, cross out the names of the uninfected partners in Round 1. There should be only two names left. One is the name of the original disease carrier. To find the original disease carrier, place a sample from his or her dropper bottle in a clean test tube, and test it with indophenol indicator. **If the procedure is followed precisely, the route of transmission can be traced easily.**

Simulating Disease Transmission *continued*

10. To show the disease transmission route, make a diagram similar to the one that follows. Show the original disease carrier and the people each disease carrier infected.

A student volunteer should construct a diagram of the route of infection on the board or overhead.

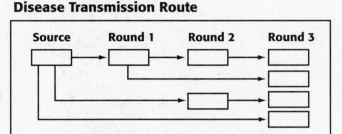

Disease Transmission Route

PART C: CLEANUP AND DISPOSAL

11. Dispose of solutions and broken glass in the designated waste containers. Do not pour chemicals down the drain unless your teacher tells you to do so.

12. Clean up your work area and all lab equipment. Return lab equipment to its proper place. Wash your hands thoroughly before you leave the lab and after you finish all work.

Analyze and Conclude

1. Interpreting Data After Round 3, how many people were "infected"? Express this number as a percentage of your class.

Answers will vary depending on class size.

2. Relating Concepts What do you think the clear fluids each student started with represent? Explain why.

The clear fluids represent pathogen-containing aerosols, or contaminated food,

water, or blood.

3. Drawing Conclusions Can someone who does not show any symptoms of a disease transmit that disease? Explain.

Yes; some pathogens must multiply in the body before symptoms appear; during

this period of time, they can be transmitted to another person.

4. Further Inquiry Write a new question about disease transmission that could be explored with another investigation.

Answers will vary. For example: Would dividing the class into two or three

quarantined groups before repeating the experiment affect disease

transmission rate?

Calculating Reaction Times

Teacher Notes

TIME REQUIRED One 50-minute class period

SKILLS ACQUIRED
Collecting Data
Constructing Materials
Experimenting
Organizing and Analyzing Data

RATING

Easy ◄— 1 —— 2 —— 3 —— 4 —► Hard

Teacher Prep–1
Student Setup–1
Concept Level–1
Cleanup–1

SCIENTIFIC METHODS

In this lab, students will:
Make Observations
Ask Questions
Test the Hypothesis
Draw Conclusions

MATERIALS

Materials for this lab can be ordered from WARD'S. Use the Lab Materials QuickList Software on the **One-Stop Planner CD-ROM** for catalog numbers and to create a customized list of materials for this lab.

SAFETY CAUTIONS

Remind students to handle the meter stick carefully and to keep it away from their face and far away from the face and eyes of their classmates. Allow students to have a practice trial. Instruct students to look only at the ruler and not at their partner. Point out that looking at their partner could distort the results of the investigation because the partner might give a clue about when he or she will drop the meterstick.

ANSWERS TO BEFORE YOU BEGIN

1. Data table should be similar to the sample on page 406.

Exploration Lab

Calculating Reaction Times

SKILLS
• Measuring
• Calculating

OBJECTIVES
• **Determine** human reaction times.
• **Design** an experiment that measures changes in reaction times.

MATERIALS
• meterstick

Before You Begin

When you want to move your hand, your brain must send a message all the way to the muscles in your arms. How long does that take? In this exercise, you will work with a partner to see how quickly you can react. In this lab, you will investigate reaction times and design an experiment to investigate influences on reaction times.

1. You will be using the data table provided to record your data.

2. Write a hypothesis about an influence on reaction times. (For example: "People who have eaten breakfast have faster reaction times than people who have not eaten breakfast.")

   Hypotheses may vary, but should be reasonable and testable. Sample answer:

   the dominant hand will have a faster reaction time than the non-dominant

   hand.

Procedure
PART A: CALCULATING REACTION TIMES

1. Sit in a chair and have a partner stand facing you while holding a meterstick in a vertical position.

2. Hold your thumb about 3 cm from your fingers near the bottom end of the stick. The meterstick should be positioned to fall between your thumb and fingers.

3. Tell your partner to let go of the meterstick without warning.

4. When your partner releases the meterstick, catch the stick by pressing your thumb and fingers together. Your partner should be ready to catch the top of the meterstick if it begins to tip over.

| Calculating Reaction Times *continued*

5. Record the number of centimeters the stick dropped before you caught it. The distance that the meterstick falls before you catch it can be used to evaluate your reaction time.

6. Repeat the procedure several times, and calculate the average number of centimeters.

7. Try this procedure with your other hand.

8. Close your eyes and have your partner say "now," when the stick is released.

9. Exchange places with your partner, and repeat the procedure.

Data Table		
Hand: trial number	**Subject 1 reaction time (s)**	**Subject 2 reaction time (s)**
Left: 1		
Left: 2		
Left: 3	**Observations in steps 2–9 will vary. Check**	
Left: average	**that students have correctly entered all**	
Right: 1	**observations in their data table.**	
Right: 2		
Right: 3		
Right: average		

PART B: DESIGNING YOUR OWN EXPERIMENT

10. Work with the members of your lab group to explore one of the hypotheses written in the **Before You Begin** section of this lab.

> **You Choose**
> As you design your experiment, decide the following:
> **a.** what hypothesis you will explore
> **c.** how you will test the hypothesis
> **d.** what the controls will be
> **e.** how many trials to perform
> **f.** what data to record in your data table

11. Write a procedure for your experiment. Make a list of all the safety precautions you will take. Have your teacher approve your procedure and safety precautions before you begin the experiment.

12. Set up your group's experiment and collect data.

| Calculating Reaction Times *continued*

Analyze and Conclude

1. **Summarizing Results** What was your fastest reaction time?

   Answers may vary.

2. **Analyzing Data** How does your reaction time when using your dominant hand compare with your reaction time when using your other hand?

   Answers may vary, but reaction times are often faster in dominant hands.

3. **Drawing Conclusions** Why may each hand have a different reaction time? Why may each person have a different reaction time? Compared to earlier trials, was the reaction time in step 8 faster, slower or the same? If the time was faster or slower, hypothesize a reason for the difference.

   Answers may vary. Reaction times calculated when the subject has closed

   eyes will probably be slower. Communication is an additional step, and

   sending and receiving the information that the stick is dropping takes time.

4. **Predicting Patterns** Compile the data gathered by each pair in your class. Can you identify any trends in the data? (For example, do males and females have the same average reaction times?)

   Answers may vary.

5. **Further Inquiry** Write a new question about reaction times that could be explored in another investigation.

   Answers may vary. Sample answer: Would training be a method of improving

   reaction time? Do students who eat breakfast have faster reaction times

   than students who skip breakfast? Can caffeine increase reaction time?

The Effect of Epinephrine on Heart Rate

Teacher Notes

TIME REQUIRED 45–60 minutes

SKILLS ACQUIRED
Collecting Data
Communicating
Designing Experiments
Experimenting
Inferring

RATING
Easy ← 1   2   3   4 → Hard

Teacher Prep–3
Student Setup–2
Concept Level–3
Cleanup–2

SCIENTIFIC METHODS

In this lab, students will:
Make Observations
Test the Hypothesis
Analyze the Results
Draw Conclusions

MATERIALS

Materials for this lab can be ordered from WARD'S. Use the Lab Materials QuickList Software on the **One-Stop Planner CD-ROM** for catalog numbers and to create a customized list of materials for this lab. Make sure all glassware is clean before beginning this experiment.

SAFETY CAUTIONS

Review all safety symbols with students before beginning the lab. Remind students to report any injuries to you. Tell students that epinephrine is toxic and is absorbed through intact skin. Remind students that chemicals used in the lab should not be touched or tasted. Spills should be reported immediately. Instruct students how to disinfect the affected surfaces following a spill. Tell students to wash hands with soap and water after the lab is finished.

TIPS AND TRICKS

When preparing solutions, mix each solution continuously for 5 min., reversing the direction of stirring every 10 to 20 seconds. To make a 0.0001 percent solution of epinephrine, dilute 10 ml of the 0.001 percent solution with distilled water to make 100 ml. To make a 0.00001 percent solution, dilute 1 ml of 0.001 percent solution with distilled water to make 100 ml. To make 0.000001 percent solution, dilute 0.1 ml of 0.001 percent solution with distilled water to make 100 ml. Prepare separate containers for the disposal of solutions and broken glass. Instruct students to get *Daphnia* from the classroom stock, and then return the *Daphnia* to a "recovery container." After the lab is finished, place *Daphnia* into a 100 ml beaker with 50 ml of *Daphnia* culture water to allow the *Daphnia* to recover. Collect all wastes containing epinephrine. To the combined liquid, add 5 ml of full-strength chlorine bleach. Stir the mixture, and then heat to boiling. Boil gently for 5 minutes. Let cool and pour down the drain.

ANSWERS TO BEFORE YOU BEGIN

1. *Epinephrine* is an amino-acid-based hormone released by the adrenal medulla in times of stress. *Heart rate* (HR) is the number of heart beats per unit of time; expressed here as beats/sec. *Threshold concentration* is the lowest concentration of a compound that stimulates a response.

Exploration Lab

The Effect of Epinephrine on Heart Rate

SKILLS

- Using scientific methods
- Graphing
- Calculating

OBJECTIVES

- **Determine** the heart rate of *Daphnia*.
- **Observe** the effect of the hormone epinephrine on heart rate in *Daphnia*.
- **Determine** the threshold concentration for the action of epinephrine on *Daphnia*.

MATERIALS

- medicine droppers
- *Daphnia*
- *Daphnia* culture water
- depression slides
- petroleum jelly
- coverslips
- compound microscope

- watch or clock with second hand
- paper towels
- 100 mL beaker
- 10 mL graduated cylinders
- epinephrine solutions (0.001%, 0.0001%, 0.00001%, and 0.000001%)

SAFETY

 CAUTION: Always wear safety goggles and a lab apron to protect your eyes and clothing.

 CAUTION: Do not touch or taste any chemicals. Know the location of the emergency shower and eyewash station and how to use them. If you get a chemical on your skin or clothing, wash it off at the sink while calling to the teacher. Notify the teacher of a spill. Spills should be cleaned up promptly, according to your teacher's directions.

 CAUTION: Glassware is fragile. Notify the teacher of broken glass or cuts. Do not clean up broken glass or spills with broken glass unless the teacher tells you to do so.

The Effect of Epinephrine on Heart Rate *continued*

Before You Begin

Epinephrine is a hormone released in response to stress. It increases blood pressure, blood glucose level, and **heart rate** (HR). The lowest concentration that stimulates a response is called the **threshold concentration**. In this lab, you will observe the effect of epinephrine on HR using the crustacean *Daphnia*. Epinephrine affects the HR of *Daphnia* and humans in similar ways.

1. Write a definition for each boldface term in the paragraph above. Use a separate sheet of paper. **Answers appear in the Teacher's Notes.**

2. You will be using the data table provided to record your data.

3. Based on the objectives for this lab, write a question you would like to explore about the action of hormones.

   Answers will vary. For example: How does epinephrine affect heart rate in _____

   Daphnia? _____

Procedure
PART A: OBSERVING HEART RATE IN *DAPHNIA*

1. **Caution: Do not touch your face while handling microorganisms.** Use a clean medicine dropper to transfer one *Daphnia* to the well of a clean depression slide. Place a dab of petroleum jelly in the well. Add a cover-slip. Observe with a compound microscope under low power.

2. Count the *Daphnia*'s heartbeats for 10 seconds. Divide this number by 10 to find the HR in beats/s. Record this number under Trial 1 in your data table. Turn off the microscope light, and wait 20 seconds. Repeat the count for Trials 2 and 3.

Sample Data:

Data Table					
Solution	**HR (beats/s) Trial 1 (A)**	**HR (beats/s) Trial 2 (B)**	**HR (beats/s) Trial 3 (C)**	**Average HR (beats/s) [(A+B+C)/3]**	**Average HR (beats/min)**
Control	24	24	25	2.4	146
0.000001%	24	25	25	2.5	148
0.00001%	25	26	25	2.5	152
0.0001%	31	32	33	3.2	192
0.001%	34	37	36	3.6	214

Name _____ Class _____ Date _____

3. After calculating the average HR in beats/s, calculate the HR in beats/min by using the following formula: HR (in beats/min) = Average HR (in beats/s) × 60 s/min.

PART B: DESIGN AN EXPERIMENT

4. Work with the members of your lab group to explore one of the questions written for step 3 of **Before You Begin.** To explore the question, design an experiment that uses the materials listed for this lab.

You Choose

As you design your experiment, decide the following:

    a. what question you will explore

    b. what hypothesis you will test

    c. how many *Daphnia* to use

    d. what your controls will be

    e. what concentrations of epinephrine to test

    f. how many trials to perform

    g. what data to record in your data table

---

5. Write a procedure for your experiment. Make a list of all the safety precautions you will take. Have your teacher approve your procedure and safety precautions before you begin the experiment.

PART C: CONDUCT YOUR EXPERIMENT

6. Put on safety goggles, gloves, and a lab apron.

7. To add a solution to a prepared slide, first place a drop of the solution at the edge of the coverslip. Then place a piece of paper towel along the opposite edge to draw the solution under the coverslip. Wait 1 minute for the solution to take effect.

8. Set up your group's experiment, and collect data. **Caution: Epinephrine is toxic and is absorbed through the skin.**

PART D: CLEANUP AND DISPOSAL

9. Dispose of solutions and broken glass in the designated waste containers. Place treated *Daphnia* in a "recovery container." Do not pour chemicals down the drain or put lab materials in the trash unless your teacher tells you to do so.

10. Clean up your work area and all lab equipment. Return lab equipment to its proper place. Wash your hands thoroughly before you leave the lab and after you finish all work.

Analyze and Conclude

1. **Summarizing Results** Use graph paper to make a graph of your group's data. Plot "Epinephrine concentration (%)" on the *x*-axis. Plot "Average heart rate (beats/min)" on the *y*-axis. **Answers will vary.**

2. **Analyzing Data** Which solutions affected the heart rate of *Daphnia*?

   The 0.001 percent and 0.0001 percent solutions should increase heart rate.

3. **Drawing Conclusions** What was the threshold concentration of epinephrine?

   Answers may vary. The threshold concentration is usually 0.0001 percent.

4. **Predicting Patterns** Based on the information you have and on your data, predict how epinephrine concentration would affect human heart rates.

   Answers will vary. It would increase them.

5. **Further Inquiry** Write a new question about hormones that could be explored with another investigation.

   Answers will vary. For example: How do other hormones affect heart rate in

   Daphnia?

Observing Embryonic Development

Teacher Notes

TIME REQUIRED One 50-minute class period

SKILLS ACQUIRED

Observing
Comparing and Contrasting
Making Drawings
Drawing Conclusions

RATING

Easy ←——— 1   2   3   4 ——→ Hard

Teacher Prep–1
Student Setup–1
Concept Level–3
Cleanup–1

SCIENTIFIC METHODS

In this lab, students will:
Make Observations
Analyze the Results
Draw Conclusions
Communicate the Results

MATERIALS

Materials for this lab can be ordered from WARD'S. Use the Lab Materials QuickList Software on the **One-Stop Planner CD-ROM** for catalog numbers and to create a customized list of materials for this lab.

SAFETY INFORMATION

Review microscope and slide safety. Tell students not to scrape the microscope objectives on the slides. If a slide breaks, remind students to be careful of its sharp edges.

TIPS AND TRICKS

Remind students that their drawings should be informative rather than artistic. If possible, have students compare the stages of sea-star development to photographs of the mammalian stages.

Name _____ Class _____ Date _____

Observing Embryonic Development

SKILLS

- Observing
- Comparing and contrasting
- Making drawings
- Drawing conclusions

OBJECTIVES

- **Identify** the stages of early animal development.
- **Describe** the changes that occur during early development.
- **Compare** the stages of human embryonic development with those of echinoderm embryonic development.

MATERIALS

- prepared slides of sea star development, including
  - unfertilized egg
  - zygote
  - 2-cell stage
  - 4-cell stage
  - 8-cell stage
  - 16-cell stage
  - 32-cell stage
  - 64-cell stage
  - blastula
  - early gastrula
  - middle gastrula
  - late gastrula
- compound light microscope
- paper and pencil

SAFETY

 CAUTION: Always wear safety goggles and a lab apron to protect your eyes and clothing.

 CAUTION: Glassware is fragile. Notify the teacher of broken glass or cuts. Do not clean up broken glass or spills with broken glass unless the teacher tells you to do so.

Before You Begin

Most members of the animal kingdom begin life as a single cell—the fertilized egg, or **zygote.** The early stages of development are quite similar in different species. Cleavage follows fertilization. During cleavage, the zygote divides many times without growing. The new cells migrate and form a hollow ball of cells called a **blastula.** The cells then begin to organize into the three primary germ layers: endoderm, mesoderm, and ectoderm. During this process, the developing organism is called a **gastrula.**

1. Write a definition for each boldface term in the preceding paragraph.

   A zygote is a fertilized egg. A blastula is a hollow ball of cells that develops

   shortly after fertilization. A gastrula is a three-layered structure that forms

   from a developing blastula.

2. Based on the objectives for this lab, write a question you would like to explore about embryonic development.

   Answers may vary.

Procedure

1. Obtain a set of prepared slides that show star eggs at different stages of development. Choose slides labeled unfertilized egg, zygote, 2-cell stage, 4-cell stage, 8-cell stage, 16-cell stage, 32-cell stage, 64-cell stage, blastula, early gastrula, middle gastrula, late gastrula, and young sea star larva. (Note: *Blastula* is the general term for the embryonic stage that results from cleavage. In mammals, a blastocyst is a modified form of the blastula.)

2. Examine each slide using a compound light microscope. Using the microscope's low-power objective first, focus on one good example of the developmental stage listed on the slide's label. Then switch to the high-power objective, and focus on the image with the fine adjustment.

3. In your lab report, draw a diagram of each developmental stage that you examine (in chronological order). Label each diagram with the name of the stage it represents and the magnification used. Record your observations as soon as they are made. Do not redraw your diagrams. Draw only what you see; lab drawings do not need to be artistic or elaborate. They should be well organized and include specific details.

4. Compare your diagrams with the diagrams of human embryonic stages shown on the next page.

Observing Embryonic Development *continued*

2-cell stage   **4-cell stage**

8-cell stage   **64-cell stage**

Blastocyst

5. Clean up your materials and wash your hands before leaving the lab.

Analyze and Conclude

1. **Summarizing Results** Compare the size of the sea star zygote with that of the blastula. At what stage does the embryo become larger than the zygote?

   The zygote and the blastula are the same size. By the time the embryo

   develops into a gastrula, it is larger than the zygote.

2. **Analyzing Data** What is the earliest stage in which all of the cells in the embryo no longer look exactly alike? How do cell shape and size change during successive stages of development?

   When the embryo reaches the gastrula stage, some cells look different from

   others. Until the gastrula stage, all cells are roughly spherical. After that,

   some become flattened, elongated, or irregularly shaped. From the first

   cleavage until the gastrula stage, the cells get progressively smaller. After

   that, some cells begin to grow.

Observing Embryonic Development *continued*

3. Drawing Conclusions From your observations of changes in cellular organization, why do you think the blastocoel (the space in the center of the hollow sphere of cells of a blastula) is important during embryonic development?

A blastocoel enables invagination (folding) to occur during gastrulation,

when the three primary germ layers form.

4. Predicting Patterns How are the symmetries of a sea star embryo and a sea star larva different from the symmetry of an adult sea star? Would you expect to see a similar change in human development? What must happen to the sea star gastrula before it becomes a mature sea star?

Both sea star embryos and larvae have bilateral symmetry, whereas the adult

has radial symmetry. No; both human embryos and adults have bilateral

symmetry. Before the gastrula becomes a mature sea star, it develops into a

larvae, which has bilateral symmetry. Because the adult has radial symmetry,

a repositioning of body organs must occur.

5. Further Inquiry How do your drawings of sea star embryonic development compare with those of human embryonic development? Based on your observations, in what ways do you think sea star embryos could be used to study early human development?

Sample answer: Sea star embryos can be used as a model of early human

development because they exhibit similar stages of cell division and differen-

tiation. Sea star eggs are a good choice for the study of early embryonic

development because they are large, numerous, and easy to culture, and

because their development is similar to that of mammals.